Lecture Note
Mathematics

Edited by A. Dold, Heidelberg and B. Eckmann, Zürich

385

Jim Douglas Jr. and Todd Dupont
The University of Chicago, Il/USA

Collocation Methods for Parabolic Equations in a Single Space Variable

Based on C^1-Piecewise-Polynomial Spaces

Springer-Verlag
Berlin · Heidelberg · New York 1974

AMS Subject Classifications (1970): Primary: 65N35, 65M99
Secondary: 65L10, 65L15

ISBN 3-540-06747-7 Springer-Verlag Berlin · Heidelberg · New York
ISBN 0-387-06747-7 Springer-Verlag New York · Heidelberg · Berlin

 Library of Congress Catalog Card Number 74-2857. Printed in Germany.

Offsetdruck: Julius Beltz, Hemsbach/Bergstr.

Preface

This manuscript presents previously unpublished research on collocation methods done over the past two years. The authors have been privileged to lecture on these topics at the University of Chicago, the Chalmers Institute of Technology, the University of Wisconsin Mathematics Research Center, l'Institut de Recherche d'Informatique et d'Automatique, and l'Université de Paris-Sud, but the manuscript does not represent a transcript of the lectures.

Problems involving more than one space variable are not considered here. We intend to discuss collocation methods for some such questions later.

The National Science Foundation supported part of this research.

Jim Douglas, Jr.
Todd Dupont

Chicago, November 1973

TABLE OF CONTENTS

CHAPTER I

GLOBAL ERROR ESTIMATES

1. Introduction. Recently the authors [8, 9] introduced a collocation method for the numerical solution of the parabolic equation

$$(1.1)\qquad \begin{gathered} c(x,t,u)\frac{\partial u}{\partial t} - a(x,t,u)\frac{\partial^2 u}{\partial x^2} - b(x,t,u,\frac{\partial u}{\partial x}) = 0\ ,\quad 0 < x < 1,\ 0 < t \le T, \\ u(x,0) = f(x)\ ,\quad 0 < x < 1, \\ u(0,t) = g_0(t),\ u(1,t) = g_1(t)\ ,\quad 0 < t \le T\ , \end{gathered}$$

based on the use of an Hermite piecewise-cubic space for approximation in the space variable. It was shown that optimal order convergence for the approximate solution to that of (1. 1) results for sufficiently smooth u for the set of collocation points chosen. In this manuscript we shall generalize and extend these results in many ways. Let

$$(1.2)\qquad \begin{gathered} 0 = x_0 < x_1 < \dots < x_M = 1,\ \delta = \{x_0, x_1, \dots, x_M\},\ h_i = x_i - x_{i-1}, \\ I_i = [x_{i-1}, x_i]\ ,\quad I = [0,1]\ . \end{gathered}$$

Let $P_r(E)$ denote the set of functions on I which are polynomials of degree not greater than r when restricted to the set E, and let

$$(1.3)\qquad \mathcal{M}_1(r,\delta) = \{v \in C^1(I) \mid v \in P_r(I_i),\ i = 1,\dots,M\}\ ,\quad r \ge 3.$$

The Hermite piecewise-cubic space is given by $\mathcal{M}_1(3,\delta)$. We shall employ as collocation points the $r-1$ points in each I_i that are the affine images of the roots of the Legendre polynomial of degree r-1. In this chapter we shall find a number of global error estimates for several collocation procedures, both continuous and discrete in the time variable. In later chapters estimates of a different nature will be derived, giving superconvergence results at the knots of the $\mathcal{M}_1(r,\delta)$-splines and indicating ways to modify the fundamental collocation method in order to get better local results.

More specifically, sections 2 and 3 of this chapter contain the definitions of two interpolation processes and a quadrature method; several technical lemmas that will be useful both in this chapter and later are proven. Section 4 contains the continuous-time collocation method based on (1.3). Existence and uniqueness of the collocation solution are proved, and a global error analysis is given. Section 5 presents two methods for finite-differencing in time the equation of section 4. First, the standard Crank-Nicolson time differencing is combined with spatial collocation. Optimal order error estimates are obtained. Then, an extrapolation is applied in the evaluation of the coefficients in the differential equation, so that the algebraic problem at each time step becomes linear even for the nonlinear equation. Again, optimal order accuracy in space and second order accuracy in time results. These methods are a special case of the methods discussed in section 6, where collocation is used in both space and time. An approximate solution is sought in a subspace constructed as the tensor product of $\mathcal{M}_1(r, \delta)$ and a C^0-piecewise polynomial space in time. This section is by far the longest (and, in our opinion, most interesting) of this chapter. The analysis of the error in these methods, both straightforward and extrapolated, is a forerunner of the basic analytic tool of Chapter II, and many of the lemmas needed in Chapter II have to be presented in this section. It is shown that optimal order accuracy in space and time occurs. The final section of the chapter is concerned with some simple extensions of the above methods.

Other references to collocation methods that relate to this manuscript include [1,6,7,17,19].

2. Two Interpolation Methods. Let $0 < \xi_1 < \xi_2 < \ldots < \xi_{r-1} < 1$ and $w_k > 0$, $k = 1, \ldots, r-1$, be the unique selection such that

$$(2.1) \qquad \int_0^1 p(x)dx = \sum_{k=1}^{r-1} p(\xi_k)w_k \ , \quad p \in P_{2r-3}(I) \ ;$$

i.e., consider the Gaussian quadrature rule [4] for $r-1$ points. Throughout, $r \geq 3$.

Lemma 2.1. There exists a unique polynomial $B_r \in P_{r+1}(I)$ such that

i) $B_r''(\xi_k) = 0$, $k = 1, \ldots, r-1$,

ii) $B_r(x) = B_r'(x) = 0$, $x = 0$ or 1,

iii) $B_r^{(r+1)}(x) \equiv 1$.

Proof. Conditions i) and iii) imply that B_r'' is a multiple of the Legendre polynomial of degree $r-1$ on the interval I [22]. Rodrigues's formula states that

$$B_r''(x) = \frac{1}{(2r-2)!} \frac{d^{r-1}}{dx^{r-1}} [x^{r-1}(x-1)^{r-1}] \ ;$$

hence

$$(2.2) \qquad B_r(x) = \frac{1}{(2r-2)!} \frac{d^{r-3}}{dx^{r-3}} [x^{r-1}(x-1)^{r-1}]$$

is the unique solution in $P_{r+1}(I)$ of i), ii), and iii).

Elementary calculus indicates that the roots of $B_r(x) = 0$ are the double roots at $x = 0$ and $x = 1$ and exactly $r-3$ simple roots at points η_j such that

(2.3) $$0 < \eta_1 < \eta_2 < \ldots < \eta_{r-3} < 1 .$$

Definition. Let $T_r : C^1(I) \to P_r(I)$ be the interpolation operator given by the conditions

i) $(T_r v)(x) = v(x)$ and $(T_r v)'(x) = v'(x)$, $x = 0$ or 1,

ii) $(T_r v)(\eta_j) = v(\eta_j)$, $j = 1, \ldots, r-3$,

for $v \in C^1(I)$.

Let

(2.4) $$C_q(x) = \frac{(r+1)!}{(r+q)!} (x - \tfrac{1}{2})^{q-1} B_r(x) , \quad q = 1, 2, \ldots .$$

It is a simple consequence of the Peano Kernel Theorem [5] that the interpolation error has the representation

(2.5) $$(v - T_r v)(x) = \sum_{q=1}^{p} v^{(r+q)}(\tfrac{1}{2}) C_q(x) + \int_0^1 k_{p,r}(x,\tau) v^{(r+p+1)}(\tau) d\tau$$

for all functions $v \in H^{r+p+1}(I)$, the Sobolev space of functions having $L^2(I)$-derivatives of order $r+p+1$ on I [18] and where the kernel is given by

$$k_{p,r}(x,\tau) = Q(x) - (T_r Q)(x) - \sum_{q=1}^{p} Q^{(r+q)}(\tfrac{1}{2}) C_q(x) ,$$

(2.6) $$Q(x) = \begin{cases} (x-\tau)^{r+p+1} , & 0 \le \tau \le x, \\ 0 & x \le \tau \le 1 . \end{cases}$$

It should be noted that the error expression (2.5) can be differentiated r+p times formally:

$$(2.7)\qquad (v - T_r v)^{(\alpha)}(x) = \sum_{q=1}^{p} v^{(r+q)}(\tfrac{1}{2})C_q^{(\alpha)}(x) + \int_0^1 \frac{\partial^\alpha k_{p,r}}{\partial x^\alpha}(x,\tau)v^{(r+p+1)}(\tau)d\tau\ ,$$

$$0 \le \alpha \le r+p\ .$$

Let

$$(2.8)\qquad {}_1\langle v, z\rangle = \sum_{j=1}^{r-1} v(\xi_j)z(\xi_j)w_j\ ,\quad {}_1|v|^2 = {}_1\langle v, v\rangle\ .$$

(Only real functions will arise.) Notice that

$$(2.9a)\qquad {}_1\langle C_q, (x-\tfrac{1}{2})^k\rangle = \text{const.}\int_0^1 (x-\tfrac{1}{2})^{q-1+k}\frac{d^{r-3}}{dx^{r-3}}(x^{r-1}(x-1)^{r-1})dx$$

$$= 0\ ,\qquad 0 \le k \le r-q-3\ .$$

Similarly,

$$(2.9b)\qquad {}_1\langle C_q^{(\ell)}, (x-\tfrac{1}{2})^k\rangle = 0\ ,\quad k-\ell \le r-q-3\ ,\quad 0 \le \ell \le 2\ .$$

These orthogonalities will play an important role in analyzing our collocation methods.

The following lemma, and a trivial generalization given immediately below, will be very important in both setting up the collocation method for (1.1) and analyzing the global error.

Lemma 2. 2. Let $v \in H^{r+3}(I)$ and set $e = v - T_r v$. Then, there exists a constant C such that

i) $\quad {}_1|e| \le C\|v^{(r+1)}\|_{L^2(I)}$,

ii) $\quad {}_1|e'| \le C\|v^{(r+1)}\|_{L^2(I)}$,

iii) $\quad {}_1|e''| \le C\|v^{(r+2)}\|_{L^2(I)}$,

iv) $\quad |{}_1\langle e', 1\rangle| \le C\|v^{(r+2)}\|_{L^2(I)}$,

v) $\quad |{}_1\langle e', 1\rangle| \le C\|v^{(r+3)}\|_{L^2(I)}$, $r \ge 4$,

vi) $\quad |{}_1\langle e'', 1\rangle| \le C\|v^{(r+3)}\|_{L^2(I)}$.

Proof. Let us prove v) and vi); the others follow similarly for error representations stopping either with the $v^{(r+1)}$ term or the $v^{(r+2)}$ term. Note that

$$e'(x) = v^{(r+1)}(\tfrac{1}{2})B_r'(x) + \frac{1}{r+2}v^{(r+2)}(\tfrac{1}{2})[(x-\tfrac{1}{2})B_r(x)]' + \int_0^1 \frac{\partial k}{\partial x}(x,\tau)v^{(r+3)}(\tau)d\tau \quad , \quad k = k_{2,r} ,$$

and that

$${}_1\langle B_r', 1\rangle = \int_0^1 B_r'(x)dx = 0 \quad , \quad r \ge 3,$$

and

$${}_1\langle [(x-\tfrac{1}{2})B_r(x)]', 1\rangle = \int_0^1 [(x-\tfrac{1}{2})B_r(x)]'dx = 0 \quad , \quad r \ge 4 .$$

Thus, v) follows from the boundedness of $\partial k/\partial x$.

Also,

$$_1\langle B_r'', 1\rangle = {_1\langle}[(x-\tfrac{1}{2})B_r(x)]'', 1\rangle = 0 \quad , \quad r \geq 3 \; ,$$

and $\partial^2 k/\partial x^2$ is bounded. Hence, vi) is valid.

We shall need to scale these results for the interval $(0, h)$. Let

$$\xi_k^h = h\xi_k \quad , \quad k = 1, \ldots, r-1 \; ,$$

$$\eta_k^h = h\eta_k \quad , \quad k = 1, \ldots, r-3 \quad ,$$

and let $T_{r,h}$ be the interpolation operator from $C^1([0,h])$ to $P_r([0,h])$ determined by replacing $(0, \eta_1, \ldots, \eta_{r-3}, 1)$ by $(0, \eta_1^h, \ldots, \eta_{r-3}^h, h)$ in the definition of T_r. Let

(2.10) $$_h\langle \alpha, \beta\rangle = \sum_{j=1}^{r-1} \alpha(\xi_j^h)\beta(\xi_j^h)w_j h, \qquad {_h|\alpha|}^2 = {_h\langle}\alpha, \alpha\rangle \; .$$

If $e(x) = (v - T_{r,h}v)(x)$ on $(0, h)$, it follows from homogeneity that

(2.11)

i) $_h|e| \leq C\|v^{(r+1)}\|_{L^2((0,h))} h^{r+1}$,

ii) $_h|e'| \leq C\|v^{(r+1)}\|_{L^2((0,h))} h^r$,

iii) $_h|e''| \leq C\|v^{(r+2)}\|_{L^2((0,h))} h^r$,

iv) $|_h\langle e', 1\rangle| \leq C\|v^{(r+2)}\|_{L^2((0,h))} h^{r+3/2}$,

v) $|_h\langle e', 1\rangle| \leq C\|v^{(r+3)}\|_{L^2((0,h))} h^{r+5/2}$, $r \geq 4$,

vi) $|_h\langle e'', 1\rangle| \leq C\|v^{(r+3)}\|_{L^2((0,h))} h^{r+3/2}$.

Let

$$(2.12)\qquad \xi_{j,k} = x_{j-1} + h_j \xi_k \;, \quad j = 1,\dots,M,\ k = 1,\dots,r-1 \;,$$

$$\eta_{j,k} = x_{j-1} + h_j \eta_k \;, \quad j = 1,\dots,M,\ k = 1,\dots,r-3 \;.$$

Then let $T_{r,\delta}$ be the interpolation operator from $C^1(I)$ to $\mathfrak{M}_1(r,\delta)$ determined by the requirements

$$(2.13)\qquad \begin{aligned} &\text{i)}\ (T_{r,\delta}v)(x_j) = v(x_j) \;,\ (T_{r,\delta}v)'(x_j) = v'(x_j) \;,\ j = 0,\dots,M \;, \\ &\text{ii)}\ (T_{r,\delta}v)(\eta_{j,k}) = v(\eta_{j,k}) \;,\ j = 1,\dots,M \;,\ k = 1,\dots,r-3 \;, \end{aligned}$$

for $v \in C^1(I)$. Clearly, $T_{r,\delta}$ is local in the sense that $T_{r,\delta}v$ is determined on I_j by v on I_j. Let

$$(2.14)\qquad \begin{aligned} &<\alpha,\beta>_j = \sum_{k=1}^{r-1} \alpha(\xi_{j,k})\beta(\xi_{j,k})w_k h_j \;, \quad |\alpha|_j^2 = <\alpha,\alpha>_j \\ &<\alpha,\beta> = \sum_{j=1}^{M} <\alpha,\beta>_j \;, \quad |\alpha|^2 = <\alpha,\alpha> \;. \end{aligned}$$

The inequalities (2.11) hold on each interval I_j with the obvious interpretations, and global forms can be written down easily. We shall, in fact, use them only on the intervals I_j .

Let us introduce another interpolation method based on the points $\xi_{j,k}$. It is not local in the sense given above.

Definition. Let v be defined at the points $0, 1$, and $\xi_{j,k}$, $j = 1,\dots,M$, $k = 1,\dots,r-1$. Then, let $S_{r,\delta}v \in \mathfrak{M}_1(r,\delta)$ be determined by the relations

$$(2.15)\quad \begin{aligned} &\text{i)}\quad (S_{r,\delta}v)(x) = v(x), \quad x = 0 \text{ or } 1, \\ &\text{ii)}\quad (S_{r,\delta}v)(\xi_{j,k}) = v(\xi_{j,k}), \quad j = 1,\dots,M,\ k = 1,\dots,r-1 . \end{aligned}$$

It is not immediately obvious that the operator $S_{r,\delta}$ exists, but its existence is implied by the following uniqueness theorem, which generalizes Lemma 4.1 of [9].

Lemma 2.3. Let $v \in \mathcal{M}_1(r,\delta)$ be such that $v(0) = v(1) = v(\xi_{j,k}) = 0$, $j = 1,\dots,M$, $k = 1,\dots,r-1$. Then, $v \equiv 0$.

Proof. Consider v on I_1. Since $v(0) = v(\xi_{1,1}) = \dots = v(\xi_{1,r-1}) = 0$ and $v \in P_r(I_1)$, either $v'(0) \neq 0$ or v vanishes identically on I_1. If v does vanish on I_1, then $v(x_1) = v'(x_1) = v(\xi_{2,1}) = \dots = v(\xi_{2,r-1}) = 0$ and v also vanishes on I_2. Obviously, the vanishing propagates across I and $v \equiv 0$. Hence, we can assume $v'(0) > 0$.

If $v'(0) > 0$, then there exist (simple) roots of $v'(x) = 0$ in $(x_0, \xi_{1,1}), (\xi_{1,1}, \xi_{1,2}), \dots, (\xi_{1,r-2}, \xi_{1,r-1})$. There are no other roots possible for v' in I_1; hence, $v(x_1)v'(x_1) > 0$. Now, consider v on I_2. Since $v \in C^1(I)$, there exist roots of $v'(x) = 0$ in $(x_1, \xi_{2,1}), (\xi_{2,1}, \xi_{2,2}), \dots, (\xi_{2,r-2}, \xi_{2,r-1})$. Consequently, $v(x_2)v'(x_2) > 0$. Again this condition propagates; thus, $v(x_M)v'(x_M) > 0$, which is a contradiction. Hence, $v \equiv 0$.

The approximation properties of $S_{r,\delta}$ will not concern us in this paper.

3. Some Quadrature Relations. There are a number of relations between the discrete inner product introduced in (2.10) and the usual $L^2(I)$ inner product.

Lemma 3.1. For $\alpha, \beta \in \mathcal{M}_1(r,\delta)$

$$-<\alpha'', \beta> = (\alpha', \beta') - \alpha'\beta\Big|_0^1 + (1-\frac{1}{r})\int_0^1 B_r''(x)^2 dx \sum_{j=1}^{M} \alpha_j^{(r)}\beta_j^{(r)} h_j^{2r-1} , \qquad (3.1)$$

where $\alpha_j^{(r)} \equiv \alpha^{(r)}(x)$, $x \in (x_{j-1}, x_j)$.

Proof. Consider the case of a single interval of length one first. Then, since $\alpha''\beta \in P_{2r-2}$ and $B_r''^{\,2} \in P_{2r-2}$, it follows that

$$(\alpha''\beta)(x) = (1-\frac{1}{r})\alpha^{(r)}\beta^{(r)}B_r''(x)^2 + p(x),$$

where $p \in P_{2r-3}$. Thus,

$$\begin{aligned} {}_1<\alpha'', \beta> &= (1-\frac{1}{r})\alpha^{(r)}\beta^{(r)}\,{}_1<B_r''(x)^2, 1> + \int_0^1 p(x)dx = \int_0^1 p(x)dx \\ &= (\alpha'', \beta) - (1-\frac{1}{r})\alpha^{(r)}\beta^{(r)}\int_0^1 B_r''(x)^2 dx . \end{aligned}$$

By homogeneity,

$$-<\alpha'', \beta> = -(\alpha'', \beta) + (1-\frac{1}{r})\int_0^1 B_r''(x)^2 dx \sum_{j=1}^{M} \alpha_j^{(r)}\beta_j^{(r)} h_j^{2r-1} .$$

Lemma 3.2. For $\alpha \in \mathcal{M}_1(r,\delta)$,

$$0 \le <\alpha', \alpha'> = (\alpha', \alpha') - \int_0^1 B_r''(x)^2 dx \sum_{j=1}^{M} \alpha_j^{(r)2} h_j^{2r-1} .$$

Proof. For $h = 1$ this follows as above; homogeneity again completes the argument.

Lemma 3.3. For $\alpha \in \mathfrak{M}^0_1(r,\delta) = \{v \in \mathfrak{M}_1(r,\delta) \mid v(0) = v(1) = 0\}$,

(3.2) $$(\alpha', \alpha') = \|\alpha\|^2_{H^1_0(I)} \leq -\langle \alpha'', \alpha \rangle \leq (2 - \frac{1}{r}) \|\alpha\|^2_{H^1_0(I)} .$$

Proof. This is an immediate consequence of Lemmas 3.1 and 3.2.

A simple calculation or a reference to Lemma 2.3 of [9] shows that

(3.3) $$\|\alpha\|^2_{H^1_0(I)} + |\alpha|^2 \geq \frac{1}{4} \|\alpha\|^2_{H^1(I)} , \quad \alpha \in H^1(I) .$$

Since $\|\cdot\|_{L^2(I)}$ is a norm on $P_r(I)$ and ${}_1|\cdot|$ is a seminorm, it is clear that

(3.4) $$|\alpha| \leq C \|\alpha\|_{L^2(I)} , \quad \alpha \in \mathfrak{M}_1(r,\delta) ,$$

where C is independent of δ. The opposite inequality can be established on $\mathfrak{M}^0_1(r,\delta)$ for uniform or quasi-uniform δ by first showing that the interpolation operator $S_{r,\delta}$ is bounded. We shall not need this result and it will be omitted. For subdivisions δ with radically varying h_i, the opposite inequality should not be expected to be true [9].

4. The Continuous Time Collocation Method. The fundamental collocation method to be studied in Chapters I-III of this paper arises from the requirement that the partial differential equation be satisfied at the points $\xi_{j,k}$, $j = 1, \ldots, M$, $k = 1, \ldots, r-1$. We seek a differentiable (in time) map $U: [0, T] \to \mathcal{M}_1(r, \delta)$ such that

$$
\begin{aligned}
&\text{i)} \quad U(x,0) - f(x) \text{ is small in a sense to be specified later,} \\
(4.1) \quad &\text{ii)} \quad \{c(U)\frac{\partial U}{\partial t} - a(U)\frac{\partial^2 U}{\partial x^2} - b(U, U_x)\}(\xi_{j,k}, t) = 0\,, \quad j = 1, \ldots, M\,,\; k = 1, \ldots, r-1\,,\; 0 < t \leq T\,, \\
&\text{iii)} \quad U(0,t) = g_0(t)\,,\; U(1,t) = g_1(t)\,, \quad 0 < t \leq T,
\end{aligned}
$$

where the writing of the dependence of the coefficients on the independent variables x and t has been suppressed. If $f \in C^1(I)$, then one very reasonable choice for $U(x, 0)$ is given by

$$
(4.2) \qquad U(x,0) = (T_{r,\delta} f)(x)\,.
$$

Other choices will be discussed in Chapter II.

We shall assume that a and c are bounded above and below by positive constants independent of their arguments and that the functions a, b, c are continuous functions of t, u, and u_x for each x. We shall also assume that a, b, c are continuously differentiable functions of u and u_x and, for simplicity, that these derivatives are all uniformly bounded; in view of the convergence estimates proved later we actually need these derivatives bounded only for arguments in a neighborhood of the values u and u_x of the solution of (1.1).

Note that it is irrelevant computationally (except for rounding error, which is not a matter of significance for the stable procedures to be introduced here) whether the coefficient of u_t or u_{xx} is normalized to one. It is, however, useful in the analysis to follow to assume

$$a(x, t, U) \equiv 1, \tag{4.3}$$

and we shall do so without loss of generality. It is also useful to introduce a continuous time discrete Galerkin procedure. Let $U:[0, T] \rightarrow \mathcal{M}_1(r, \delta)$ satisfy (4.1.i) with the same $U(x, 0)$, (4.1.iii), and

$$< c(U)\frac{\partial U}{\partial t} - \frac{\partial^2 U}{\partial x^2} - b(U, U_x), z> = 0 , \quad z \in \mathcal{M}_1(r, \delta), \ 0 < t \leq T. \tag{4.4}$$

Let us emphasize that the orthogonality is with respect to the discrete inner product (2.14).

<u>Lemma 4.1.</u> The collocation method (4.1) and the discrete Galerkin method (4.4) each possess a unique solution for $0 < t \leq T$; moreover, these solutions are identical if the processes are started from the same initial values.

<u>Proof.</u> It is clear that any solution of (4.1) is a solution of (4.4). Thus, it is sufficient to prove existence for (4.1) and uniqueness for (4.4). Existence for (4.1) follows from Lemma 2.3, since it implies that matrix

generated by the time derivative term is nonsingular for any choice of the basis for $\mathcal{M}_1(r, \delta)$. Uniqueness for solutions of (4.4) also is implied by Lemma 2.3, since the matrix generated by the time-derivative term in (4.4) must be nonsingular since $c(x, t, U)$ is bounded below by a positive constant.

Let us turn to the convergence of the solution U of (4.4) to the solution u of (1.1). Set

$$W(x,t) = (T_{r,\delta}u(\cdot\,,t))(x)\,, \quad 0 \le t \le T\,; \tag{4.5}$$

this assumes that u is sufficiently smooth that $u(x_i, t)$ and $\partial u/\partial x(x_i, t)$ exist as functions of t. We shall assume noticeably more smoothness on u before we are through. Let

$$\zeta = u - U, \quad \eta = u - W\,, \quad \nu = W - U\,. \tag{4.6}$$

Note that $\nu(0,t) = \nu(1,t) = 0$. Then, (4.1), (4.4) and the argument of §5 of [9] imply that

$$\begin{aligned} \langle c(U)\nu_t - \nu_{xx}, v\rangle &= \langle -c_3W_t\nu - c\eta_t - c_3u_t\eta + b_3\eta + b_3\nu + b_4\nu_x, v\rangle \\ &\quad + \langle \eta_{xx}, v\rangle + \langle b(W, u_x) - b(W, W_x), v\rangle \end{aligned} \tag{4.7}$$

for all $v \in \mathcal{M}_1(r, \delta)$, where we have assumed b and c differentiable with respect to u and u_x and c_j is an evaluation of the derivative of c with respect to its j-th argument. Then, the choice $v = \nu_t$ for the test function and a repetition of the argument of §5 of [9] show that (with $m = \inf c(x, t, v)$)

$$m\int_0^t |\nu_t|^2 d\tau - \langle \nu_{xx}, \nu \rangle \Big|_0^t + |\nu|^2 \Big|_0^t$$

(4.8)
$$\leq C\int_0^t (|\nu|^2 + |\nu_x|^2) d\tau + C\int_0^t (|\eta|^2 + |\eta_t|^2) d\tau$$
$$+ 2\int_0^t \langle \eta_{xx}, \nu_t \rangle d\tau + 2\int_0^t \langle b(W, u_x) - b(W, W_x), \nu_t \rangle d\tau.$$

Now,

$$\int_0^t \langle \eta_{xx}, \nu_t \rangle d\tau = \langle \eta_{xx}, \nu \rangle \Big|_0^t - \int_0^t \langle \eta_{xxt}, \nu \rangle d\tau .$$

For any fixed time t,

$$\langle \eta_{xx}, \nu \rangle = \sum_{i=1}^{M} \langle \eta_{xx}, \nu \rangle_i$$
$$= \sum_{i=1}^{M} \langle \eta_{xx}, \bar{\nu}_i \rangle_i + \sum_{i=1}^{M} \langle \eta_{xx}, \nu - \bar{\nu}_i \rangle_i ,$$

where $\bar{\nu}_i = h_i^{-1} \langle \nu, 1 \rangle_i$. The inequality (2.11.vi) implies that

$$|\langle \eta_{xx}, \bar{\nu}_i \rangle_i| \leq C \left\| \frac{\partial^{r+3} u}{\partial x^{r+3}} \right\|_{L^2(I_i)} h_i^{r+3/2} \cdot h_i^{-1} |\nu|_i h_i^{1/2}$$
$$= C \left\| \frac{\partial^{r+3} u}{\partial x^{r+3}} \right\|_{L^2(I_i)} h_i^{r+1} |\nu|_i .$$

A trivial case of the Poincaré inequality shows that

$$|\langle \eta_{xx}, \nu - \bar{\nu}_i \rangle_i| \leq |\eta_{xx}|_i |\nu - \bar{\nu}_i|_i$$
$$\leq C \left\| \frac{\partial^{r+2} u}{\partial x^{r+2}} \right\|_{L^2(I_i)} h_i^r \|\nu_x\|_{L^2(I_i)} h_i$$
$$= C \left\| \frac{\partial^{r+2} u}{\partial x^{r+2}} \right\|_{L^2(I_i)} \|\nu_x\|_{L^2(I_i)} h_i^{r+1} .$$

Hence,

$$|<\eta_{xx}, \nu>| \leq C \sum_{i=1}^{M} \|u\|_{H^{r+3}(I_i)} (\|\nu_x\|_{L^2(I_i)} + |\nu|_i) h_i^{r+1}$$

$$\leq \frac{1}{4}(\|\nu_x\|^2_{L^2(I)} + |\nu|^2) + C \sum_{i=1}^{M} \|u\|^2_{H^{r+3}(I_i)} h_i^{2r+2} .$$

Lemma 3.3 can be employed to obtain

$$|<\eta_{xx}, \nu>| \leq \frac{1}{4}(-<\nu_{xx}, \nu> + |\nu|^2) + C \sum_{i=1}^{M} \|u\|^2_{H^{r+3}(I_i)} h_i^{2r+2}. \tag{4.9}$$

Exactly the same argument can be used to show that

$$|\int_0^t <\eta_{xxt}, \nu> d\tau| \leq \frac{1}{4}\int_0^t (-<\nu_{xx}, \nu> + |\nu|^2) d\tau$$
$$+ C \sum_{i=1}^{M} h_i^{2r+2} \int_0^t \|u_t\|^2_{H^{r+3}(I_i)} d\tau . \tag{4.10}$$

Assume that b has bounded third derivatives in the neighborhood of the solution u. An argument paralleling that of §5 of [9] shows that

$$\int_0^t <b(W, u_x) - b(W, W_x), \nu_t> d\tau$$
$$= <b(W, u_x) - b(W, W_x), \nu> \Big|_0^t$$
$$- \int_0^t < \frac{\partial}{\partial t}[b(W, u_x) - b(W, W_x)], \nu> d\tau$$

and

$$|<b(W, u_x) - b(W, W_x), \nu>| \leq \frac{1}{4}(-<\nu_{xx}, \nu> + |\nu|^2)$$
$$+ C \sum_{i=1}^{M} \|u\|^2_{H^{r+2}(I_i)} h_i^{2r+2} , \quad 0 \leq t \leq T,$$

(4.11)

$$\left|\int_0^t < \frac{\partial}{\partial t}[b(W,u_x) - b(W,W_x)], \nu > d\tau\right| \le \frac{1}{4}\int_0^t (-<\nu_{xx}, \nu> + |\nu|^2) d\tau$$

$$+ C\sum_{i=1}^{M} h_i^{2r+2}\int_0^t (\|u\|^2_{H^{r+2}(I_i)} + \|u_t\|^2_{H^{r+2}(I_i)}) d\tau .$$

It then follows from (4.8)-(4.11), Lemmas 3.2 and 3.3, and the Gronwall lemma that

$$|\nu_t|^2_{L^2(0,T)} + \|\nu\|^2_{L^\infty(0,T;H^1(I))}$$

(4.12)

$$\le C[\|\nu(0)\|^2_{H^1(I)} + \sum_{i=1}^{M} (\|u\|^2_{L^\infty(0,T;H^{r+3}(I_i))} + \|u_t\|^2_{L^2(0,T;H^{r+3}(I_i))}) h_i^{2r+2}] .$$

In particular,

(4.13)

$$\|\nu\|_{L^\infty(I\times[0,T])} \le C[\|\nu(0)\|_{H^1(I)} + \{\sum_{i=1}^{M} (\|u\|^2_{L^\infty(0,T;H^{r+3}(I_i))} + \|u_t\|^2_{L^2(0,T;H^{r+3}(I_i))}) h_i^{2r+2}\}^{1/2}] .$$

It is easy to see, using (2.5) with $p = 0$, that

(4.14)

$$\|\eta\|_{L^\infty(I\times[0,T])} \le C\{\sum_{i=1}^{M} \|u\|^2_{L^\infty(0,T;H^{r+2}(I_i))} h_i^{2r+2}\}^{1/2} ;$$

hence,

$$\|u-U\|_{L^{\infty}(I\times[0,T])} \leq C\Big[\|\nu(0)\|_{H^1(I)} + \Big\{\sum_{i=1}^{M}\Big(\|u\|^2_{L^{\infty}(0,T;H^{r+3}(I_i))}$$

(4.15)

$$+ \|u_t\|^2_{L^2(0,T;H^{r+3}(I_i))}\Big)h_i^{2r+2}\Big\}^{1/2}\Big] .$$

The natural choice of the initial condition $U(x,0)$ is $(T_{r,\delta}f)(x) = W(x,0)$ so that $\nu(0) = 0$. We shall see in Chapter II that, if we wish in addition to obtain the possible higher order accuracy (for $r \geq 4$) at the knots x_i, there are better choices if u is sufficiently smooth; however, we shall take $U(0) = T_{r,\delta}f$ for the moment.

<u>Theorem 4.1.</u> Let the coefficients a and c in the differential equation (1.1) have bounded third derivatives in a neighborhood of the closure of the set $\{(x,t,u(x,t)) \mid x \in I,\ t \in [0,T]\}$, where u is the solution of (1.1), and let b have bounded third derivatives in a neighborhood of the closure of $\{(x,t,u(x,t),u_x(x,t)) \mid x \in I,\ t \in [0,T]\}$. Assume that

$$u \in L^{\infty}(0,T;H^{r+3}(I_i)),\quad u_t \in L^2(0,T;H^{r+3}(I_i)),\quad i = 1,\dots,M.$$

Let $U(0) = T_{r,\delta}f$. Then there exists a unique solution U of the collocation equations (4.1), and

(4.16)
$$\|u-U\|_{L^{\infty}(I\times[0,T])}$$
$$\leq C\Big\{\sum_{i=1}^{M}\Big(\|u\|^2_{L^{\infty}(0,T;H^{r+3}(I_i))} + \|u_t\|^2_{L^{\infty}(0,T;H^{r+3}(I_i))}\Big)h_i^{2r+2}\Big\}^{1/2}.$$

In the special case that $h_i \leq h$, $i = 1, \ldots, M$, then the error estimate can reduce to

$$(4.17) \quad \|u-U\|_{L^\infty(I\times[0,T])} \leq C(\|u\|_{L^\infty(0,T;H^{r+3}(I))} + \|u_t\|_{L^2(0,T;H^{r+3}(I))})h^{r+1}.$$

The estimate (4.16) contains more information than the simplified estimate (4.17). If the coefficients are piecewise smooth in x and if knots are introduced at each (fixed in time) point of discontinuity of a coefficient, it can be the case that the solution is smooth on each $I_i \times [0, T]$. If so, the optimal order global convergence takes place; see §7 for indications of methods to treat such problems.

5. Discretization in Time by Finite Differences. The practical calculation of an approximate solution of (1.1) by the collocation method indicated by (4.1) requires that the continuous time process be discretized in time. In this section several finite difference discretizations will be introduced and analyzed.

Let

$$U^n = U^n(x) = U(x, t_n) , \quad t_n = n\Delta t , \quad \Delta t = T/N,$$

$$U^{n+\alpha} = \alpha U^{n+1} + (1-\alpha)U^n , \quad 0 < \alpha < 1 ,$$

$$d_t U^n = (U^{n+1} - U^n)/\Delta t .$$

Then the Crank-Nicolson collocation method corresponding to (4.1) is the following. Find a map $U: \{t_0, t_1, \ldots, t_N\} \to \mathcal{M}_1(r,\delta)$ such that

$$\text{i)} \quad U^0 - f \quad \text{is small,}$$

$$\text{ii)} \quad \{c(U^{n+1/2})d_t U^n - a(U^{n+1/2})U_{xx}^{n+1/2} - b(U^{n+1/2}, U_x^{n+1/2})\} (\xi_{ij}) = 0 ,$$

$$(5.1) \qquad i = 1, \ldots, M, \; j = 1, \ldots, r-1, \; n = 0, \ldots, N-1,$$

$$\text{iii)} \quad U^n(0) = g_0(t_n) , \; U^n(1) = g_1(t_n) , \quad n = 0, \ldots, N .$$

The proper interpretation of $c(U^{n+1/2})(\xi_{ij})$ is

$$c(U^{n+1/2})(\xi_{ij}) = c(\xi_{ij}, t_{n+1/2}, U^{n+1/2}(\xi_{ij})) ,$$

and the other coefficients are to be evaluated similarly.

We shall continue using the normalization that $a(x, t, u) \equiv 1$, still without loss of generality. It is easily seen that a unique solution of (5.1) exists for c and b continuously differentiable, c bounded below, and Δt

sufficiently small. Moreover, (5.1.ii) is equivalent to

(5.1.ii') $$<c(U^{n+1/2})d_t U^n - U_{xx}^{n+1/2} - b(U^{n+1/2}, U_x^{n+1/2}), z> = 0 ,$$

$$z \in \mathfrak{M}_1(r, \delta) , \quad 0 \le n < N .$$

The analysis of the convergence of the solution of (5.1) to that of (1.1) can be made to parallel the argument given for the continuous time case. Let $W = T_{r\delta} u$, $0 \le t \le T$, as before. It is easy to see that

$$<c(W^{n+1/2})d_t W^n - W_{xx}^{n+1/2}, z>$$

$$= <c(u^{n+1/2})d_t u^n - u_{xx}^{n+1/2}, z> - <c_3 d_t W^n \cdot \eta^{n+1/2}, z>$$

$$+ <-c(u^{n+1/2})d_t \eta^n + \eta_{xx}^{n+1/2}, z>$$

(5.2) $$= <c(u)u_t - u_{xx} - b(u, u_x), z>(t_{n+1/2})$$

$$+ <[c(u^{n+1/2}) - c(u(t_{n+1/2}))]d_t u^n, z>$$

$$+ <c(u(t_{n+1/2}))[d_t u^n - u_t(t_{n+1/2})] - [u_{xx}^{n+1/2} - u_{xx}(t_{n+1/2})], z>$$

$$+ <b(u, u_x), z>(t_{n+1/2}) - <c_3 d_t W^n \cdot \eta^{n+1/2} + c(u^{n+1/2})d_t \eta^n - \eta_{xx}^{n+1/2}, z>,$$

where $\eta = u - W$ and $\nu = W - U$. It follows that

$$<c(U^{n+1/2})d_t \nu^n - \nu_{xx}^{n+1/2}, z>$$

(5.3) $$= <-c_3 d_t W^n \cdot \nu^{n+1/2}, z> + <\delta^n, z>$$

$$+ < b(u^{n+1/2}, u_x^{n+1/2}) - b(U^{n+1/2}, U_x^{n+1/2}), z >$$

$$- <c_3 d_t W^n \cdot \eta^{n+1/2} + c(u^{n+1/2})d_t \eta^n - \eta_{xx}^{n+1/2}, z > ,$$

where

$$\delta^n = [c(u^{n+1/2}) - c(u(t_{n+1/2}))]d_t u^n + c(u(t_{n+1/2}))[d_t u^n - u_t(t_{n+1/2})]$$

(5.4)

$$- [u_{xx}^{n+1/2} - u_{xx}(t_{n+1/2})] + [b(u, u_x)(t_{n+1/2}) - b(u^{n+1/2}, u_x^{n+1/2})] .$$

Let

(5.5) $$\|u\|_{\alpha,\beta} = \sup\{ \left| \frac{\partial^{k+j} u}{\partial x^k \partial t^j}(x,t) \right| \Big| x \in I_i,\ 0 \leq t \leq T,\ 1 \leq i \leq M,\ 0 \leq k \leq \alpha,\ 0 \leq j \leq \beta \} .$$

If b and c have bounded derivatives in the neighborhood of the solution u of (1.1), then

(5.6) $$|\delta^n(\xi_{ij})| \leq C(\|u\|_{2,2}, \|u\|_{0,3})(\Delta t)^2 .$$

Choose as test function $z = d_t \nu^n$. Then,

$$m|d_t\nu^n|^2 - \langle \nu_{xx}^{n+1/2}, d_t\nu^n \rangle$$

$$\leq \frac{m}{4}|d_t\nu^n|^2 + C[|\nu^{n+1/2}|^2 + |\eta^{n+1/2}|^2 + |d_t\eta^n|^2 + (\Delta t)^4]$$

$$+ \langle \eta_{xx}^{n+1/2}, d_t\nu^n \rangle + \langle b(u^{n+1/2}, u_x^{n+1/2}) - b(U^{n+1/2}, U_x^{n+1/2}), d_t\nu^n \rangle .$$

Lemma 3.1 implies that

$$\langle \nu_{xx}^{n+1/2}, d_t\nu^n \rangle = \frac{1}{2} d_t \langle \nu_{xx}^n, \nu^n \rangle .$$

Also,

$$d_t|\nu^n|^2 = \langle d_t\nu^n, \nu^n + \nu^{n+1} \rangle \leq \frac{m}{4}|d_t\nu^n|^2 + C(|\nu^n|^2 + |\nu^{n+1}|^2) .$$

As in the continuous time case,

$$< b(u^{n+1/2}, u_x^{n+1/2}) - b(U^{n+1/2}, U_x^{n+1/2}), d_t\nu^n >$$

$$= < b(W^{n+1/2}, u_x^{n+1/2}) - b(W^{n+1/2}, W_x^{n+1/2}), d_t\nu^n >$$

$$+ \psi_1^n \frac{m}{8} |d_t\nu^n|^2 + \psi_2^n C[|\eta^{n+1/2}|^2 + |\nu^{n+1/2}|^2 + |\nu_x^{n+1/2}|^2],$$

$$|\psi_k^n| \le 1 \ , \ k = 1, 2 .$$

Hence,

$$\frac{m}{2}|d_t\nu^n|^2 - \frac{1}{2}d_t< \nu_{xx}^n, \nu^n> + \frac{1}{2}d_t|\nu^n|^2$$

$$(5.7) \qquad \le C[|\nu^n|^2 + |\nu^{n+1}|^2 + |\nu_x^{n+1/2}|^2 + |\eta^{n+1/2}|^2 + |d_t\eta^n|^2 + (\Delta t)^4]$$

$$+ < b(W^{n+1/2}, u_x^{n+1/2}) - b(W^{n+1/2}, W_x^{n+1/2}), d_t\nu^n > + < \eta_{xx}^{n+1/2}, d_t\nu^n > .$$

Sum in time:

$$m \sum_{k=0}^{n} |d_t\nu^n|^2 \Delta t - < \nu_{xx}^{n+1}, \nu^{n+1} > + |\nu^{n+1}|^2$$

$$(5.8) \qquad \begin{aligned} &\le - < \nu_{xx}^0, \nu^0 > + |\nu^0|^2 + C[(\Delta t)^4 + \sum_{k=0}^{n+1} (|\nu^k|^2 + |\nu_x^{k+1}|^2)\Delta t \\ &+ \sum_{k=0}^{n} (|\eta^{k+1/2}|^2 + |d_t\eta^k|^2)\Delta t \\ &+ \sum_{k=0}^{n} < \eta_{xx}^{k+1/2} + b(W^{k+1/2}, u_x^{k+1/2}) - b(W^{k+1/2}, W_x^{k+1/2}), d_t\nu^k > \Delta t] . \end{aligned}$$

Let $\partial_t\nu^n = (\nu^{n+1} - \nu^{n-1})/(2\Delta t)$. Then,

$$\sum_{k=0}^{n} < \eta_{xx}^{k+1/2}, d_t\nu^k > \Delta t = < \eta_{xx}^{n+1}, \nu^{n+1} > - < \eta_{xx}^0, \nu^0 >$$

$$- \sum_{k=1}^{n} < \partial_t\eta_{xx}^k, \nu^k > \Delta t \ ,$$

and it is easy to see [9] that

$$\Big|\sum_{k=0}^{n} \langle \eta_{xx}^{k+1/2}, d_t \nu^k \rangle \Delta t\Big| \le \frac{1}{4}\big(-\langle \nu_{xx}^{n+1}, \nu^{n+1}\rangle + |\nu^{n+1}|^2\big)$$

$$+ C\|\nu(0)\|^2_{H^1(I)} + C\sum_{i=1}^{M} h_i^{2r+2}\big(\|u^{n+1}\|^2_{H^{r+3}(I_i)} + \|u^0\|^2_{H^{r+3}(I_i)}\big)$$

$$+ \frac{1}{4}\sum_{k=0}^{n}\big(-\langle \nu_{xx}^k, \nu^k\rangle + |\nu^k|^2\big)\Delta t$$

$$+ C\sum_{i=1}^{M} h_i^{2r+2}\int_0^{t^{n+1}} \|u_t\|^2_{H^{r+3}(I_i)}\, d\tau \;.$$

The term involving the difference of the b-evaluations can also be summed by parts in time and then treated by the argument of [9] used in the continuous time case above to show that

$$\Big|\sum_{k=0}^{n} \langle b(W^{k+1/2}, u_x^{k+1/2}) - b(W^{k+1/2}, W_x^{k+1/2}), d_t \nu^k \rangle \Delta t\Big|$$

$$\le \frac{1}{4}\big(-\langle \nu_{xx}^{n+1}, \nu^{n+1}\rangle + |\nu^{n+1}|^2\big) + C\|\nu(0)\|^2_{H^1(I)}$$

$$+ \frac{1}{4}\sum_{k=0}^{n}\big(-\langle \nu_{xx}^k, \nu^k\rangle + |\nu^k|^2\big)\Delta t +$$

$$+ C\sum_{i=1}^{M} h_i^{2r+2}\big(\|u^{n+1}\|^2_{H^{r+2}(I_i)} + \|u^0\|^2_{H^{r+2}(I_i)}\big)$$

$$+ \sum_{i=1}^{M} h_i^{2r+2}\int_0^{t^{n+1}} \big(\|u\|^2_{H^{r+2}(I_i)} + \|u_t\|^2_{H^{r+2}(I_i)}\big)\, d\tau \;.$$

Thus, if $U^0 = T_{r,\delta} f$, so that $\nu(0) = 0$,

$$\sum_{k=0}^{N-1} |d_t \nu^k|^2 \Delta t + \max_{0 \le k \le N} \|\nu^k\|^2_{H^1(I)}$$

(5.9) $$\le C\Big[\sum_{i=1}^{M} \big(\|u\|^2_{L^\infty(0,T;H^{r+3}(I_i))} + \|u_t\|^2_{L^2(0,T;H^{r+3}(I_i))}\big) h_i^{2r+2} + (\Delta t)^4\Big],$$

$$C = C(\|u\|_{2,2}, \|u\|_{0,3}) .$$

Therefore,

$$\max_{0 \le k \le N} \|(u-U)^k\|_{L^\infty(I)} \le C(\|u\|_{2,2}, \|u\|_{0,3})[(\Delta t)^2$$

(5.10) $$+ \{\sum_{i=1}^{M} (\|u\|^2_{L^\infty(0,T;H^{r+3}(I_i))} + \|u_t\|^2_{L^2(0,T;H^{r+3}(I_i))}) h_i^{2r+2}\}^{1/2}]$$

In particular, if $h_i \le h$ and $\|u\|_{2,2} + \|u\|_{1,r+3} < \infty$, then

$$\max_{0 \le k \le N} \|(u-U)^k\|_{L^\infty} \le C[(\Delta t)^2 + h^{r+1}] .$$

Theorem 5.1. Let the hypothesis of Theorem 4.1 hold and in addition let $\|u\|_{2,2}$ and $\|u\|_{0,3}$ be finite. For Δt sufficiently small, there exists a unique solution of the Crank-Nicolson collocation equation (5.1) and the error between u and U satisfies the inequality (5.10).

The equation (5.1) is algebraically nonlinear if the differential equation is. It is possible to extrapolate the values of u in the coefficients in order to avoid nonlinearity in the algebraic problem. Let

(5.11) $$\breve{U}^n = \frac{3}{2} U^n - \frac{1}{2} U^{n-1}$$

and consider the extrapolated Crank-Nicolson collocation method:

$$\{c(\tilde{U}^n)d_tU^n - U_{xx}^{n+1/2} - b(\tilde{U}^n, \tilde{U}_x^n)\}(\xi_{ij}) = 0,\tag{5.12}$$

$$i = 1,\ldots,M,\ j = 1,\ldots,r-1,\ n = 1,\ldots,N-1,$$

subject to the boundary conditions (5.1.iii) and the initial condition $U^0 = T_{r,\delta}f$. It is also necessary to obtain the vector U^1 by some start-up procedure before using the equation (5.12) to advance the approximate solution in time. Note that (5.12) represents a linear system for the coefficients with respect to any basis for $\mathcal{M}_1(r,\delta)$ of the vector U^{n+1}. It is clear that there exists a unique solution to (5.12), given U^0 and U^1.

The analysis of the convergence of the solution of (5.12) to that of (1.1) is a perturbation of the argument given for the unmodified Crank-Nicolson procedure. It follows from (5.2) and (5.12) that

$$\begin{aligned}
&< c(\tilde{U}^n)d_t\nu^n - \nu_{xx}^{n+1/2}, d_t\nu^n > \\
&\quad = < [c(\tilde{U}^n) - c(W^{n+1/2})]d_tW^n, d_t\nu^n > \\
&\qquad + < [c(u^{n+1/2}) - c(u(t_{n+1/2}))]d_tu^n + c(u(t_{n+1/2}))[d_tu^n - u_t(t_{n+1/2})], d_t\nu^n > \\
&\qquad - < u_{xx}^{n+1/2} - u_{xx}(t_{n+1/2}), d_t\nu^n > \\
&\qquad + < b(u,u_x)(t_{n+1/2}) - b(\tilde{U}^n, \tilde{U}_x^n), d_t\nu^n > \\
&\qquad - < c_3 d_tW^n\cdot\eta^{n+1/2} + c(u^{n+1/2})d_t\eta^n - \eta_{xx}^{n+1/2}, d_t\nu^n > .
\end{aligned}\tag{5.13}$$

Now,

$$c(\tilde{U}^n) - c(W^{n+1/2}) = [c(\tilde{U}^n) - c(\check{W}^n)] + [c(\check{W}^n) - c(W^{n+1/2})]$$

$$= c_3 \tilde{\nu}^n + c_3(\tilde{W}^n - W^{n+1/2})$$

$$= c_3 \tilde{\nu}^n - \frac{1}{2} c_3 (W^{n+1} - 2W^n + W^{n-1}) ,$$

and

$$\| W^{n+1} - 2W^n + W^{n-1} \|_{0,0} \leq C \| u \|_{1,2} (\Delta t)^2 .$$

Also,

$$|< b(u, u_x)(t_{n+1/2}) - b(\tilde{U}^n, \tilde{U}^n_x), d_t \nu^n > - < b(\tilde{W}^n, \tilde{u}^n_x) - b(\tilde{W}^n, \tilde{W}^n_x), d_t \nu^n >|$$

$$\leq C(u)[(\Delta t)^4 + |\tilde{\eta}^n|^2 + |\tilde{\nu}^n|^2 + |\tilde{\nu}^n_x|^2] + \frac{m}{8} |d_t \nu^n|^2 ,$$

where

$$C(u) \leq C(\| u \|_{1,3}) .$$

It then follows that

$$\frac{m}{2} |d_t \nu^n|^2 - \frac{1}{2} d_t < \nu^n_{xx}, \nu^n > + \frac{1}{2} |\nu^n|^2$$

$$\leq C[\sum_{k=n-1}^{n+1} (|\nu^k|^2 + |\nu^k_x|^2) + |d_t \eta^n|^2 + \sum_{k=n-1}^{n+1} |\eta^k|^2 + (\Delta t)^4]$$

(5.14)

$$+ < \eta^{n+1/2}_{xx}, d_t \nu^n > + < b(\tilde{W}^n, \tilde{u}^n_x) - b(\tilde{W}^n, \tilde{W}^n_x), d_t \nu^n > .$$

The summations by parts, now starting at $n = 1$, can be carried out as before ; it can be shown that

$$\sum_{k=1}^{N-1} |d_t \nu^n|^2 \Delta t + \max_{2 \le k \le N} \|\nu^k\|^2_{H^1(I)}$$

(5.15)
$$\le C(\|u\|_{2,2}, \|u\|_{1,3})[(\Delta t)^4 + \|\nu^1\|^2_{H^1(I)} + \sum_{i=1}^{M} (\|u\|^2_{L^\infty(0,T;H^{r+3}(I_i))} + \|u_t\|^2_{L^2(0,T;H^{r+3}(I_i))})h_i^{2r+2}].$$

If U^1 is determined by iterating out the Crank-Nicolson equation (5.1) for this single step (and it is a trivial programming problem to take the code for (5.12) and make it do a step of (5.1) first), then $\| \nu(0)\|_{H^1(I)}$ can be estimated by (5.10) with T replaced by Δt. Then,

(5.16)
$$\max_{1 \le k \le N} \|(u-U)^k\|_{L^\infty(I)} \le C(\|u\|_{2,2}, \|u\|_{1,3})[(\Delta t)^2 + \{\sum_{i=1}^{M}(\|u\|^2_{L^\infty(0,T;H^{r+3}(I_i))} + \|u_t\|^2_{L^2(0,T;H^{r+3}(I_i))})h_i^{2r+2}\}^{1/2}].$$

<u>Theorem 5.2.</u> Let the hypotheses of Theorem 5.1 hold and assume that $\|u\|_{1,3}$ is finite. Let U^n denote the solution of the extrapolated Crank-Nicolson equation (5.12) starting from the initial values $U^0 = T_{r,\delta}f$ and U^1, U^1 being the solution of (5.1). Then the error between u and U satisfies the inequality (5.16).

Note that the second-order accuracy in the time step has been preserved. The error estimate above can be extended to include the case in which predictor-corrector modification of (5.1) with the same preservation of order of accuracy. It is also the case that $\tilde{U}^n$ could have been defined by

$$\tilde{U}^n = 2U^{n-1/2} - U^{n-3/2} ,$$

and almost exactly the same estimates result; it is necessary to obtain both U^1 and U^2 by using (5.1) or a predictor-corrector. Probably this extrapolation is a bit smoother than (5.11), but it requires keeping slightly more information on hand.

The equation (5.1) and its extrapolated version (5.12) are special cases of the full collocation procedures to be studied in the next section.

6. Discretization in Time by Collocation. Higher-order accuracy in the time-discretization can be obtained by the use of a collocation method in time combined with the collocation method using $\mathcal{M}_1(r,\delta)$ in space. Let ϵ denote the partition of $J = [0,T]$ given by

$$\epsilon = \{0 = t_0, t_1, \ldots, t_N = T\}, \quad \Delta t_k = t_k - t_{k-1} > 0, \quad J_k = [t_{k-1}, t_k].$$

It is natural for parabolic equations to require only continuity of the approximate solution in time at the knots t_k, so that the solution can be calculated on each time interval independently. See also [17,6] for the initial value problem for ordinary differential equations. Let

$$(6.1) \qquad \mathcal{M}_0(s,\epsilon) = \{v \in C^0([0,T]) \mid v \in P_s(J_k), \quad k = 1,\ldots,N\},$$

and let

$$(6.2) \qquad \mathcal{M} = \mathcal{M}_1(r,\delta) \otimes \mathcal{M}_0(s,\epsilon);$$

i.e., a basis for $\mathcal{M}$ is given by the (finite) collection of products of the form $w(x)v(t)$, w an element of a basis for $\mathcal{M}_1(r,\delta)$ and v an element of a basis for $\mathcal{M}_0(x,\epsilon)$. Let

$$(6.3) \qquad 0 < \tau_1 < \tau_2 < \ldots < \tau_s < 1$$

be the Gauss quadrature points in $[0,1]$ such that

$$\int_0^1 p(t)dt = \sum_{\ell=1}^{s} p(\tau_\ell)w_\ell^*, \quad p \in P_{2s-1}([0,1]),$$

and set

$$(6.4) \qquad \tau_{k\ell} = t_{k-1} + \Delta t_k \tau_\ell, \quad k = 1,\ldots,N, \quad \ell = 1,\ldots,s.$$

Let

$$(6.5)\qquad B_s^*(t) = \frac{1}{(2s)!}\frac{d^{s-1}}{dt^{s-1}}(t^s(t-1)^s) \quad , \quad 0 \le t \le 1 .$$

Then

$$\frac{d}{dt}B_s^*(\tau_\ell) = 0 \, , \quad \ell = 1,\ldots,s \, ,$$

$$(6.6)\qquad B_s^*(0) = B_s^*(1) = 0 \, ,$$

$$\frac{d^{s+1}}{dt^{s+1}} B_s^*(t) \equiv 1 \, .$$

Moreover, there exist simple zeros of $B_s^*(t) = 0$ at points σ_q such that

$$(6.7)\qquad 0 = \sigma_0 < \sigma_1 < \ldots < \sigma_s = 1 .$$

Let $T_s^*: C^0([0,1]) \to P_s([0,1])$ denote the standard Lagrangian interpolation process defined by

$$(6.8)\qquad (T_s^* v)(\sigma_q) = v(\sigma_q) \, , \quad q = 0,\ldots,s .$$

If

$$C_j^*(t) = \frac{(s+1)!}{(s+j)!}\,(t-\tfrac{1}{2})^{j-1} B_s^*(t) \, ,$$

then

$$(6.9)\qquad (v - T_s^* v)(t) = \sum_{j=1}^{q} v^{(s+j)}(\tfrac{1}{2}) C_j^*(t) + \int_0^1 K_{qs}^*(t,\tau) v^{(s+q+1)}(\tau)\, d\tau$$

for any function $v \in H^{s+q+1}([0,1])$. Let

$$(6.10)\qquad {}_1[v,z] = \sum_{\ell=1}^{s} v(\tau_\ell) z(\tau_\ell) w_\ell^* \, , \quad {}_1[v]^2 = {}_1[v,v]$$

denote a discrete inner product and norm, respectively, associated with the points τ_ℓ. Then ,

$$(6.11)\qquad {}_1\left[\frac{d^\alpha}{dt^\alpha} B_s^*, t^\beta\right] = \begin{cases} \left(\frac{d^\alpha}{dt^\alpha} B_s^*, t^\beta\right), & \beta \le \alpha + s - 2, \\ 0, & 0 \le \alpha \le 1,\ \beta \le \alpha + s - 2. \end{cases}$$

It follows that

$$\text{i)}\qquad {}_1[v - T_s^* v] \le C\|v^{(s+1)}\|_{L^2([0,1])},$$

$$(6.12)\ \text{ii)}\qquad {}_1\left[\frac{d}{dt}(v - T_s^* v)\right] \le C\|v^{(s+2)}\|_{L^2([0,1])},$$

$$\text{iii)}\qquad \left|{}_1\left[\frac{d^\alpha}{dt^\alpha}(v - T_s^* v), t^\beta\right]\right| \le C\|v^{(s+\ell+1)}\|_{L^2([0,1])}, \quad \alpha \le 1,\ \alpha - \beta \ge \ell - s + 1.$$

Define an interpolation process $T_{s,\epsilon}^* : C^0([0,T]) \to \mathcal{M}_0(s,\epsilon)$ by the relations

$$(6.13)\qquad (T_{s,\epsilon}^* v)(\sigma_{kq}) = v(\sigma_{kq}), \quad k = 1, \ldots, N, \quad q = 0, \ldots, s,$$

where

$$\sigma_{kq} = t_{k-1} + \Delta t_k \sigma_q .$$

<u>Definition.</u> Let $\mathcal{J} = \mathcal{J}_{r,\delta;s,\epsilon} = T_{r,\delta} \otimes T_{s,\epsilon}^* : C^1(I \times [0,T]) \to \mathcal{M}$ be the interpolation process determined by the relations

$$\text{i)}\quad \left(\frac{\partial^\alpha}{\partial x^\alpha} \mathcal{J} u\right)(x_i, \sigma_{kq}) = \frac{\partial^\alpha u}{\partial x^\alpha}(x_i, \sigma_{kq}), \quad \alpha = 0 \text{ or } 1,\ i = 0, \ldots, M,$$

$$(6.14)\qquad k = 1, \ldots, N, \quad q = 0, \ldots, s,$$

$$\text{ii)}\quad (\mathcal{J} u)(\eta_{ij}, \sigma_{kq}) = u(\eta_{ij}, \sigma_{kq}), \quad i = 1, \ldots, M,\ j = 1, \ldots, r-1,$$

$$k = 1, \ldots, N, \quad q = 0, \ldots, s.$$

Note that the operator $I - \mathcal{J}$ can be written in the form

$$(6.15)\qquad I - \mathcal{J} = (I - T_{r,\delta}) \otimes I + I \otimes (I - T^*_{s,\epsilon}) - (I - T_{r,\delta}) \otimes (I - T^*_{s,\epsilon}).$$

With the notation that has been introduced above, it is possible to formulate a full collocation approximation of (1.1). Let $U \in \mathcal{M}$ be determined by

$$(6.16)\qquad \begin{aligned} &\text{i)} \quad U(x,0) = (T_{r,\delta} f)(x), \\ &\text{ii)} \quad U(0,t) = (T^*_{s,\epsilon} g_0)(t), \quad U(1,t) = (T^*_{s,\epsilon} g_1)(t), \\ &\text{iii)} \quad \{ c(U)U_t - a(U)U_{xx} - b(U, U_x) \}(\xi_{ij}, \tau_{k\ell}) = 0, \\ &\qquad\qquad i = 1, \ldots, M, \ j = 1, \ldots, r-1, \\ &\qquad\qquad k = 1, \ldots, N, \ \ell = 1, \ldots, s. \end{aligned}$$

It is easy to see that (6. 16) is local in time. First, the interpolations of the initial and boundary conditions are both straight-forward and local. Then, (6. 16, iii) used for all (i, j, ℓ) for fixed k allows the advancement of the solution over the time interval J_k. If $s = 1$, then (6. 16) is the Crank-Nicolson scheme (5. 1).

The analysis of the convergence of (6. 16) will be given first for the linear problem

$$(6.17)\qquad \begin{aligned} &\text{i)} \quad u_t - a(x,t)u_{xx} - b(x,t)u_x - e(x,t)u = g(x,t), \quad x \in I, \ 0 < t \leq T, \\ &\text{ii)} \quad u(x,0) = f(x), \quad x \in I, \\ &\text{iii)} \quad u(0,t) = u(1,t) = 0, \quad 0 < t \leq T. \end{aligned}$$

Note that the coefficient of the time derivative term has been normalized to one rather than that of t..e second derivative with respect to x; obviously, no loss of generality is caused by this normalization. Let

$$L = \frac{\partial}{\partial t} - a\frac{\partial^2}{\partial x^2} - b\frac{\partial}{\partial x} - e$$

and consider $L(I - \mathcal{J})u$ at the collocation points $(\xi_{ij}, \tau_{k\ell})$. For convenience, let $h_i = h$ and $\Delta t_k = \Delta t$ be taken constant.

The term $\frac{\partial}{\partial t}(I - \mathcal{J})u$ can be broken into three terms by use of (6 . 15). First,

$$\{\frac{\partial}{\partial t}(I - T_{r,\delta}) \otimes Iu\}(\xi_{ij}, t) = (I - T_{r,\delta})\frac{\partial u}{\partial t}(\xi_{ij}, t)$$

$$= O(\|\frac{\partial u}{\partial t}\|_{r+1,0} h^{r+1}),$$

by (2. 4), used for $q = 0$, and homogeneity in the interval length. Next

$$\{\frac{\partial}{\partial t} I \otimes (I - T^*_{s,\epsilon})u\}(x, \tau_{k\ell}) = \{\frac{\partial}{\partial t}(I - T^*_{s,\epsilon})u\}(x, \tau_{k\ell})$$

$$= O(\|u\|_{0,s+2}(\Delta t)^{s+1}),$$

by (6. 6), (6. 9), $q = 1$, and homogeneity. Also, it can be shown that

$$\{\frac{\partial}{\partial t}(I-T_{r,\delta}) \otimes (I-T^*_{s,\epsilon})u\}(\xi_{ij}, \tau_{k\ell}) = O(\|u\|_{r-1,2} h^{r-1}\Delta t)$$

as follows. Since $T_{r,\delta}$ is obviously exact on $\mathcal{M}_1(r-2, \delta)$, it follows from the Peano Kernel Theorem, applied on subintervals I_i, that, for $x \in I_i$,

$$\{(I - T_{r,\delta}) \otimes Iu\}(x,t) = h^{r-1}\int_0^1 K_1(\frac{x-x_{i-1}}{h}, \alpha)\frac{\partial^{r-1}u}{\partial x^{r-1}}(x_{i-1} + \frac{\alpha}{h}, t)\, d\alpha .$$

Similarly, for $s \geq 1$, $I-T^*_{s,\epsilon}$ is exact on $\mathcal{M}_0(1,\epsilon)$. Hence, for $t \epsilon J_k$,

$$\{I \otimes (I-T^*_{s,\epsilon})u\}(x,t) = (\Delta t)^2 \int_0^1 K_2(\frac{t-t_{k-1}}{\Delta t}, \beta) \frac{\partial^2 u}{\partial t^2}(x, t_{k-1} + \frac{\beta}{\Delta t}) d\beta ,$$

and the integral can be differentiated formally once with respect to t.

Thus,

$$\{\frac{\partial}{\partial t}(I-T_{r,\delta}) \otimes (I-T^*_{s,\epsilon})u\}(x,t)$$

$$= h^{r-1} \Delta t \int_0^1 \int_0^1 K_1(\frac{x-x_{i-1}}{h}, \alpha) \frac{\partial K_2}{\partial t}(\frac{t-t_{k-1}}{\Delta t}, \beta) \frac{\partial^{r+1} u}{\partial x^{r-1} \partial t^2}(x_{i-1} + \frac{\alpha}{h}, t_{k-1} + \frac{\beta}{\Delta t}) d\alpha d\beta$$

for $x \epsilon I_i$ and $t \epsilon J_k$, and the estimate above is proved. The motivation for the choice of the exponents will be given shortly.

The treatment of the term $\frac{\partial^2}{\partial x^2}(I-\mathcal{J})u$ must be done somewhat more carefully. Again, the three terms generated by (6.15) could be considered, but it is slightly more convenient to combine the first and third ones to get a term of the form $\frac{\partial^2}{\partial x^2}(I-T_{r,\delta}) \otimes T^*_{s,\epsilon} u$. It is convenient to use the usual representation [4] of Lagrangian interpolation for this term. Let

$$d_q \epsilon P_s([0,1]), \quad d_q(\sigma_p) = \delta_{pq}, \quad p, q = 0, \ldots, s .$$

Then,

$$(T^*_{s\epsilon} u)(x,t) = \sum_{q=0}^{s} d_q(\frac{t-t_{k-1}}{\Delta t}) u(x, \sigma_{kq}), \quad t \epsilon J_k .$$

Lemma 2.1, (2.4), and homogeneity imply that

$$\{\frac{\partial^2}{\partial x^2}(I-T_{r\delta})\otimes T^*_{s\epsilon}u\}(\xi_{ij},t) = h^r\sum_{q=0}^{s} d_q\left(\frac{t-t_{k-1}}{\Delta t}\right)\frac{\partial^{r+2}u}{\partial x^{r+2}}(x_{i-1/2},\sigma_{kq})C_2''(\xi_j)$$

$$+ O(\|u\|_{r+3,0}h^{r+1}) \;, \quad t \in J_k \;.$$

It follows from (6.9) that

$$\{\frac{\partial^2}{\partial x^2}I\otimes(I-T^*_{s\epsilon})u\}(x,t) = O(\|u\|_{2,s+1}(\Delta t)^{s+1}) \;.$$

Thus, if $a(x,t)$ has bounded first derivatives,

$$\{a\frac{\partial^2}{\partial x^2}(I-\mathcal{J})u\}(\xi_{ij},t) = h^r\sum_{q=0}^{s} d_q\left(\frac{t-t_{k-1}}{\Delta t}\right)a(x_{i-1/2},\sigma_{kq})\frac{\partial^{r+2}u}{\partial x^{r+2}}(x_{i-1/2},\sigma_{kq})C_2''(\xi_j)$$

$$+ O(\|u\|_{r+3,0}h^{r+1} + \|u\|_{r+2,0}h^r\Delta t + \|u\|_{2,s+1}(\Delta t)^{s+1})$$

for $t \in J_k$, with the $h^r\Delta t$ term arising from replacing $a(x_{i-1/2},t)$ by $a(x_{i-1/2},\sigma_{kq})$ after first replacing $a(\xi_{ij},t)$ by $a(x_{i-1/2},t)$ and absorbing this error in the h^{r+1} term that already existed.

A similar study shows that, for $t \in J_k$,

$$\{b\frac{\partial}{\partial x}(I-\mathcal{J})u\}(\xi_{ij},t) = h^r\sum_{q=0}^{s} d_q\left(\frac{t-t_{k-1}}{\Delta t}\right)b(x_{i-1/2},\sigma_{kq})\frac{\partial^{r+1}u}{\partial x^{r+1}}(x_{i-1/2},\sigma_{kq})C_1'(\xi_j)$$

$$+ O(\|u\|_{r+2,0}h^{r+1} + \|u\|_{r+1,0}h^r\Delta t + \|u\|_{1,s+1}(\Delta t)^{s+1}).$$

Also,

$$\{e(I-\mathcal{J})u\}(\xi_{ij},t) = O(\|u\|_{r+1,0}h^{r+1} + \|u\|_{0,s+1}(\Delta t)^{s+1}) \;.$$

These estimates can be combined to give the expression

$$\{L(I-\mathcal{J})u\}(\xi_{ij},\tau_{k\ell}) = -h^r \sum_{q=0}^{s} d_q\left(\frac{t-t_{k-1}}{\Delta t}\right)\Big[\Big(a\frac{\partial^{r+2}u}{\partial x^{r+2}}\Big)(x_{i-1/2},\sigma_{kq})C_2''(\xi_j)$$

$$(6.18) \qquad + \Big(b\frac{\partial^{r+1}u}{\partial x^{r+1}}\Big)(x_{i-1/2},\sigma_{kq})C_1'(\xi_j)\Big]$$

$$+ O((\|u\|_{r+3,0} + \|u\|_{r+1,1})h^{r+1} + \|u\|_{r-1,2}h^{r-1}\Delta t$$

$$+ (\|u\|_{0,s+2} + \|u\|_{2,s+1})(\Delta t)^{s+1} + \|u\|_{r+2,0}h^r\Delta t).$$

From this point on the argument is a preview of the method of analysis that will be developed in great detail in Chapter II of this paper. It was shown in §2 that

$$(6.19) \qquad {}_1\langle C_1', 1\rangle = {}_1\langle C_2'', 1\rangle = 0.$$

These orthogonalities will be used to construct a modification $\hat{u}$ of $\mathcal{J}u$ such that the leading term in $L(u-\mathcal{J}u)$ is annihilated at the collocation points. This construction will be given here for the case $r \geq 4$; a more complicated argument will be indicated for $r = 3$ in an Appendix in Chapter II using only the orthogonalities of (6.19). For $r \geq 4$, it is also the case that

$$(6.20) \qquad {}_1\langle C_1', x\rangle = {}_1\langle C_2'', x\rangle = 0.$$

The following lemma will be very useful in the construction.

Lemma 6.1. Let ${}_1\langle F, x^\alpha \rangle = 0$ for $\alpha = 0$ and 1 and let $r \geq 4$. Then there exists a unique $D \in P_r(I)$ such that

i) $D''(\xi_j) = F(\xi_j)$, $j = 1, \ldots, r-1$,

ii) $D(0) = D'(0) = D(1) = D'(1) = 0$.

Moreover, ${}_1\langle D', 1 \rangle = 0$, and

$$\|D\|_{L^\infty(I)} + \|D'\|_{L^\infty(I)} \leq C \max_j |F(\xi_j)| .$$

Proof. $D'' \in P_{r-2}(I)$ is uniquely determined by i). The additional conditions $D(0) = D'(0) = 0$ determine $D \in P_r(I)$ exactly. Now,

$$D'(1) = \int_0^1 D''(x)dx = {}_1\langle D'', 1 \rangle = {}_1\langle F, 1 \rangle = 0$$

and

$$D(1) = \int_0^1 (1-x)D''(x)dx = {}_1\langle F, 1-x \rangle = 0 .$$

Also, ${}_1\langle D', 1 \rangle = D(1) = 0$. Finally, the inequality follows from the linearity of the map $F \to D$ and the finite dimensionality of the spaces.

Lemma 6.1 can be extended, essentially by scaling, as below.

Lemma 6.2. Let $\langle F, x^\alpha \rangle_i = 0$, $\alpha = 0$ and 1, $i = 1, \ldots, M$, and let $r \geq 4$. Then there exists a unique $D \in \mathcal{M}_1(r, \delta)$ such that

i) $h^2 D''(\xi_{ij}) = F(\xi_{ij})$, $i = 1, \ldots, M$, $j = 1, \ldots, r-1$,

ii) $D(x_i) = D'(x_i) = 0$, $i = 0, \ldots, M$.

Moreover,

$$\|D\|_{L^\infty(I)} + h\|D'\|_{L^\infty(I)} \leq C \max_{i,j} |F(\xi_{ij})| .$$

<u>Proof.</u> Let $D(x) = E_i\left(\frac{x-x_{i-1}}{h}\right)$ on I_i, where $E''_i(\xi_j) = F(\xi_{ij})$ and $E_i(0) = E'_i(0) = E_i(1) = E'_i(1) = 0$. Lemma 6.1 implies the existence and uniqueness of each E_i. Thus, D exists and is unique and satisfies the inequality of the lemma.

Lemma 6.2 can be employed to show that there exists $D(\cdot, t) \in \mathcal{M}_1(r, \delta)$, $r \geq 4$, such that, for $t \in J_k$,

$$h^2 D_{xx}(\xi_{ij}, t) = \sum_{q=0}^{s} d_q\left(\frac{t-t_{k-1}}{\Delta t}\right)\left[\frac{\partial^{r+2} u}{\partial x^{r+2}}(x_{i-1/2}, \sigma_{kq})C''_2(\xi_j)\right.$$

(6.21)
$$\left. + \left(\frac{b}{a}\frac{\partial^{r+1} u}{\partial x^{r+1}}\right)(x_{i-1/2}, \sigma_{kq})C'_1(\xi_j)\right], \quad i = 1, \ldots, M, \quad j = 1, \ldots, r-1,$$

$$D(x_i, t) = D_x(x_i, t) = 0 \quad , \quad i = 0, \ldots, M.$$

Note that $D \in \mathcal{M}$ and $D_t(\cdot, t) \in \mathcal{M}_1(r, \delta)$ and that, by Lemma 6.2,

(6.22)
$$\|D_t\|_{0,0} \leq C\|u\|_{r+2,1} .$$

Also,

(6.23)
$$h^2 a(\xi_{ij}, t)D_{xx}(\xi_{ij}, t) = a(x_{i-1/2}, t)\frac{\partial^{r+2} u}{\partial x^{r+2}}(x_{i-1/2}, t)C''_2(\xi_j) + b(x_{i-1/2}, t)\frac{\partial^{r+1} u}{\partial x^{r+1}}(x_{i-1/2}, t)C'_1(\xi_j) + O(\|u\|_{r+2,1}\Delta t).$$

Let

(6.24)
$$\hat{u}(x, t) = (\mathcal{J} u)(x, t) - h^{r+2}D(x, t).$$

Then,

$$L(u-\hat{u})(\xi_{ij},\tau_{k\ell}) = O(\|u\|_{r+3,0} + \|u\|_{r+1,1} + h\|u\|_{r+2,1})h^{r+1}$$

$$+ (\|u\|_{r-1,2} + h\|u\|_{r+2,1})h^{r-1}\Delta t$$

(6.25)
$$+ (\|u\|_{0,s+2} + \|u\|_{2,s+1})(\Delta t)^{s+1})$$

$$\equiv -\rho(\xi_{ij},\tau_{k\ell}) .$$

Let $\mathcal{M} \ni \nu = U-\hat{u} = (U-u) + (u-\hat{u})$. Since $L(u-U)(\xi_{ij},\tau_{k\ell}) = 0$,

$$< -\nu_t + a\nu_{xx} + b\nu_x + e\nu, \nu_{xx} >(\tau_{k\ell}) = < \rho , \nu_{xx} > (\tau_{k\ell}) .$$

For $a(x,t) \geq m > 0$ and b and e bounded,

$$-\frac{d}{dt} < \nu_{xx}, \nu > + m|\nu_{xx}|^2 \leq C[|\rho|^2 + |\nu_x|^2 + |\nu|^2] , \quad t = \tau_{k\ell} .$$

Since $\frac{d}{dt} < \nu_{xx}, \nu > \in P_{2s-1}(J_k)$, Gaussian quadrature is exact and

$$-< \nu_{xx}, \nu > \Big|_{t_{k-1}}^{t_k} + m \sum_{\ell=1}^{s} |\nu_{xx}(\tau_{k\ell})|^2 w^*_\ell \Delta t$$

$$\leq C \sum_{\ell=1}^{s} \{|\rho(\tau_{k\ell})|^2 + |\nu_x(\tau_{k\ell})|^2 + |\nu(\tau_{k\ell})|^2\} w^*_\ell \Delta t .$$

It follows from the equation $L\nu = -\rho$ that

$$|\nu_t|^2 \leq C[|\nu_{xx}|^2 + |\nu_x|^2 + |\nu|^2 + |\rho|^2] .$$

Consequently, there exists a positive γ such that

$$-< \nu_{xx}, \nu > \Big|_{t_{k-1}}^{t_k} + \frac{m}{2} \sum_{\ell=1}^{s} |\nu_{xx}(\tau_{k\ell})|^2 w^*_\ell \Delta t + \gamma \sum_{\ell=1}^{s} |\nu_t(\tau_{k\ell})|^2 w^*_\ell \Delta t$$

(6.26)
$$\leq C \sum_{\ell=1}^{s} \{|\rho(\tau_{k\ell})|^2 + |\nu_x(\tau_{k\ell})|^2 + |\nu(\tau_{k\ell})|^2\} w^*_\ell \Delta t .$$

It is helpful to indicate some lemmas at this point.

Lemma 6.3. If $\varphi \in H^1([0,\alpha])$ and $\beta \in [0,\alpha]$, then

$$\|\varphi\|^2_{L^2([0,\alpha])} \leq 2\alpha\varphi(\beta)^2 + 4\alpha^2 \int_0^\alpha \varphi'(x)^2 dx .$$

Proof. Since $\varphi(x)^2 = \varphi(\beta)^2 + \int_\beta^x [\varphi(\xi)^2]' d\xi$,

$$\int_0^\alpha \varphi(x)^2 dx = \alpha\varphi(\beta)^2 + 2\{ -\int_0^\beta \xi\varphi(\xi)\varphi'(\xi) d\xi + \int_\beta^\alpha (\alpha-\xi)\varphi(\xi)\varphi'(\xi) d\xi \}$$

$$\leq \alpha\varphi(\beta)^2 + \frac{1}{2}\int_0^\alpha \varphi(\xi)^2 d\xi + 2\alpha^2 \int_0^\alpha \varphi'(\xi)^2 d\xi .$$

Corollary 6.4. If $\varphi \in H^1(I)$, then

$$\|\varphi\|^2_{L^2(I)} \leq 2|\varphi|^2 + 4h^2 \|\varphi\|^2_{H^1_0(I)} .$$

Corollary 6.5. If $\varphi \in \mathcal{M}_1(r,\delta) \otimes \mathcal{M}_0(s,\epsilon)$, then

$$\int_{J_k} |\varphi(t)|^2 dt \leq 2|\varphi(t_{k-1})|^2 \Delta t + 4(\Delta t)^2 \int_{J_k} |\varphi_t(t)|^2 dt .$$

Lemma 3.3, the compactness of the injection of $H^2(I)$ into $H^1(I)$, and Corollary 6.4 imply that

$$C|v_x|^2 \leq C\|v_x\|^2_{L^2(I)} \leq \frac{m}{8}\|v_{xx}\|^2_{L^2(I)} + C\|v\|^2_{L^2(I)}$$

$$\leq \frac{m}{8}|v_{xx}|^2 + C|v|^2 + Ch^2|v_{xx}|^2 ;$$

hence, for h sufficiently small,

$$C|\nu_x|^2 \leq \frac{m}{4}|\nu_{xx}|^2 + C|\nu|^2 .$$

Thus, with of course a new constant,

$$-\langle \nu_{xx}, \nu \rangle \Big|_{t_{k-1}}^{t_k} + \frac{m}{4}\sum_{\ell=1}^{s} |\nu_{xx}(\tau_{k\ell})|^2 w_\ell^* \Delta t + \gamma \sum_{\ell=1}^{s} |\nu_t(\tau_{k\ell})|^2 w_\ell^* \Delta t$$

$$\leq C \sum_{\ell=1}^{s} \{|\rho(\tau_{k\ell})|^2 + |\nu(\tau_{k\ell})|^2\} w_\ell^* \Delta t .$$

Since any $p \in P_s(I)$ can be written in the form $p = p_0 + p_1$, $p_0 \in P_{s-1}$, $p_1(\tau_\ell) = 0$, $\ell = 1,\dots,s$, it follows easily that

$$_1[p]^2 \leq \int_0^1 p(\tau)^2 d\tau , \quad p \in P_s(I).$$

Hence,

$$\sum_{\ell=1}^{s} |\nu(\tau_{k\ell})|^2 w_\ell^* \Delta t \leq \int_{J_k} |\nu(t)|^2 dt .$$

It follows from Corollary 6.5 and the above that

$$C\sum_{\ell=1}^{s} |\nu(\tau_{k\ell})|^2 w_\ell^* \Delta t \leq 2C|\nu(t_{k-1})|^2 \Delta t + 4C(\Delta t)^2 \int_{J_k} |\nu_t(t)|^2 dt$$

$$= 2C|\nu(t_{k-1})|^2 \Delta t + 4C(\Delta t)^2 \sum_{\ell=1}^{s} |\nu_t(\tau_{k\ell})|^2 w_\ell^* \Delta t$$

since $|\nu_t|^2 \in P_{2s-2}(J_k)$ and the quadrature is exact. The embedding (3.4) and that of $H_0^1(I)$ into $L^2(I)$ and Lemma 3.3 show that

$$|\nu|^2 \leq C\|\nu\|^2_{L^2(I)} \leq C\|\nu_x\|^2_{L^2(I)} \leq -C\langle \nu_{xx}, \nu \rangle .$$

Thus, if $4C(\Delta t)^2 \le \frac{1}{2}\gamma$,

$$-<\nu_{xx},\nu>\Big|_{t_{k-1}}^{t_k} + \frac{m}{4}\sum_{\ell=1}^{s} |\nu_{xx}(\tau_{k\ell})|^2 w_\ell^* \Delta t + \frac{\gamma}{2}\sum_{\ell=1}^{s} |\nu_t(\tau_{k\ell})|^2 w_\ell^* \Delta t$$

$$\le C\{-<\nu_{xx},\nu>(t_{k-1})\Delta t + \sum_{\ell=1}^{s} |\rho(\tau_{k\ell})|^2 w_\ell^* \Delta t\} .$$

Finally, the Gronwall lemma and Lemma 3. 3 imply that

$$\max_{0\le t_k\le T} \|\nu(t_k)\|^2_{H^1(I)} + \sum_{k=1}^{N}\sum_{\ell=1}^{s} \|\nu_{xx}(\tau_{k\ell})\|^2_{L^2(I)} w_\ell^* \Delta t + \int_0^T |\nu_t(t)|^2 dt$$

(6. 27)
$$\le C\{\|\nu(0)\|^2_{H^1(I)} + \sum_{k=1}^{N}\sum_{\ell=1}^{s} |\rho(\tau_{k\ell})|^2 w_\ell^* \Delta t\} .$$

Since $\nu(0) = (U-\hat{u})(0) = h^{r+2}D(\cdot\,,0)$, it follows from Lemma 6. 2 that

(6. 28)
$$\|\nu(0)\|^2_{H^1(I)} \le C\|u\|^2_{r+2,0}\, h^{2r+2} .$$

Hence

$$\max_{0\le t_k\le T} \|\nu(t_k)\|^2_{H^1(I)} + \sum_{k=1}^{N}\sum_{\ell=1}^{s} \|\nu_{xx}(\tau_{k\ell})\|^2_{L^2(I)} w_\ell^* \Delta t + \int_0^T |\nu_t(t)|^2 dt$$

(6. 29)
$$\le G(h,\Delta t)^2 ,$$

where

$$G(h,\Delta t) = C[(\|u\|_{r+3,0} + \|u\|_{r+1,1} + h\|u\|_{r+2,1})h^{r+1}$$

(6.30)
$$+ (\|u\|_{r-1,2} + h\|u\|_{r+2,1})h^{r-1}\Delta t$$

$$+ (\|u\|_{0,s+2} + \|u\|_{2,s+1})(\Delta t)^{s+1}] .$$

In particular,

$$\max_{0 \le t_k \le T} \|\nu(t_k)\|_{L^\infty(I)} \le G(h, \Delta t) , \tag{6.31}$$

without constraint between h and Δt. It is natural, however, to relate r and s and h and Δt. First, the norms appearing in (6.30) are not independent. Clearly, one time differentiation is essentially equivalent to two spatial differentiations. Thus, the leading terms in the expression (6.30) require approximately the same smoothness on the solution u if

$$r = 2s + 1 . \tag{6.32}$$

This is not possible for even r, but s can be chosen so that the difference is at most one. It also seems reasonable to make the three error terms of the same order of magnitude:

$$h^{r+1} \doteq h^{r-1} \Delta t \doteq (\Delta t)^{s+1} . \tag{6.33}$$

If (6.32) holds, then (6.33) leads to the choice

$$0 < c_1 \le \frac{\Delta t}{h^2} \le c_2 < \infty . \tag{6.34}$$

If (6.32) fails to hold, it is still reasonable to expect a choice of the form

$$0 < c_1 \le \frac{\Delta t}{h} + \frac{h^3}{\Delta t} \le c_2 < \infty . \tag{6.35}$$

If (6.35) holds, it is possible to obtain an L^∞-estimate on u-U for $(x,t) \in I \times [0,T]$ of optimal order in h and Δt by means of the following lemma.

Lemma 6.6. Let $h_i = h$ and $\Delta t_k = \Delta t$. Then the norms

$$\|\varphi\|_{L^\infty(I\times[0,T])} \quad \text{and}$$

$$\max_{\substack{i=1,\dots,M\\ k=1,\dots,N}} \{\|\varphi(\cdot,t_k)\|^2_{L^\infty(I_i)} + \frac{h^3}{\Delta t}\sum_{\ell=1}^{s} |\varphi_{xx}|^2_i(\tau_{k\ell})w^*_\ell\,\Delta t + \frac{\Delta t}{h}\int_{J_k} |\varphi_t|^2_i(\tau)d\tau\}^{1/2}$$

are equivalent on $\mathcal{M}_1(r,\delta)\times\mathcal{M}_0(s,\epsilon)$ for $r\geq 3$. In particular,

$$\|\varphi\|_{L^\infty(I\times[0,T])} \leq C\{\max_{k=0,\dots,N} \|\varphi(\cdot,t_k)\|^2_{L^\infty(I)} + \frac{h^3}{\Delta t}\sum_{k=1}^{N}\sum_{\ell=1}^{s} |\varphi_{xx}|^2(\tau_{k\ell})w^*_\ell\,\Delta t + \frac{\Delta t}{h}\int_0^T |\varphi_t|^2(\tau)d\tau\}^{1/2}.$$

Proof. First consider the unit square and $\psi\in P_r([0,1])\otimes P_s([0,1])$. Let us show that

$$\psi(x,1) = 0, \quad x\in I,$$

and

$$\sum_{\ell=1}^{s} |\psi_{xx}(\tau_\ell)|^2 w^*_\ell + \int_0^1 |\psi_t(t)|^2\,dt = 0$$

imply $\psi\equiv 0$. It follows immediately that $\psi(x,\tau_\ell)$ is linear in x for each $\ell=1,\dots,s$, and that $\psi(\xi_j,t)$ is constant for each $j=1,\dots,r-1$. Since $\psi(x,1)=0$, then

$$\psi(\xi_j,\tau_\ell) = \psi(0,\tau_\ell) = \psi(1,\tau_\ell) = \psi(\xi_j,1) = 0.$$

Thus, $\psi \equiv 0$. Hence, there exist positive constants c' and c'' such that

$$c'\|\psi\|^2_{L^\infty([0,1]^2)} \le \|\psi(\cdot,1)\|^2_{L^\infty(I)} + \sum_{\ell=1}^{s} |\psi_{xx}(\tau_\ell)|^2 w_\ell^* + \int_0^1 |\psi_t(t)|^2 dt$$

$$\le c''\|\psi\|^2_{L^\infty([0,1]^2)} .$$

Homogeneity gives the immediate implication that

$$c'\|\psi\|^2_{L^\infty([0,h]\times[0,\Delta t])} \le \|\psi(\cdot,\Delta t)\|^2_{L^\infty([0,h])} + \frac{h^3}{\Delta t}\sum_{\ell=1}^{s} h|\psi_{xx}(\tau_{1\ell})|^2 w_\ell^* \Delta t$$

$$(6.36)\qquad + \frac{\Delta t}{h}\int_0^{\Delta t} h|\psi_t(t)|^2 dt$$

$$\le c''\|\psi\|^2_{L^\infty([0,h]\times[0,\Delta t])}$$

for $\psi \in P_r([0,h]) \otimes P_s([0,\Delta t])$. Thus, the lemma has been proved.

Assume that (6.35) is valid. Then application of Lemma 6.6 gives the estimate

$$(6.37)\qquad \|v\|_{L^\infty(I\times[0,T])} \le CG(h,\Delta t) .$$

Now

$$(6.38)\qquad \|u-\hat{u}\|_{L^\infty(I\times[0,T])} \le C[(\|u\|_{r+1,0} + h\|u\|_{r+2,0})h^{r+1} + \|u\|_{0,s+1}(\Delta t)^{s+1}]$$

for $r \geq 4$; for $r = 3$, the h multiplying $\|u\|_{r+2,0}$ will not appear.

Thus,

$$\|u-U\|_{L^\infty(I \times [0,T])} \leq CG(h, \Delta t) \ . \tag{6.39}$$

The results derived above can be summarized as follows.

Theorem 6.1. Let u be the solution of the linear problem (6.17) and let $U \in \mathcal{M}_1(r, \delta) \otimes \mathcal{M}_0(x, \epsilon)$ be the solution of the full collocation equations (6.16) corresponding to (6.17). Let $G(h, \Delta t)$ be defined by (6.30) and assume that the norms in (6.30) are finite for u. Assume that $a, b, e \in C^1(I \times [0, T])$. Then,

$$\max_{k=1,\ldots,N} \|(u-U)(t_k)\|_{L^\infty(I)} \leq CG(h, \Delta t) \ .$$

If, in addition,

$$\frac{\Delta t}{h} + \frac{h^3}{\Delta t} \leq c < \infty \ ,$$

then

$$\|u-U\|_{L^\infty(I \times [0,T])} \leq CG(h, \Delta t) \ .$$

An estimate for $(u-U)_x$ can be obtained from (6.36) and the remainder of the proof of Lemma 6.6. Since $\|\psi_x\|_{L^\infty([0,1]^2)}$ is a semi-norm on $P_r([0,1]) \otimes P_s([0,1])$, it follows that

$$h^2 \|\psi_x\|^2_{L^\infty([0,h]\times[0,\Delta t])} \le C\{\|\psi(\cdot,\Delta t)\|^2_{L^\infty([0,h])} + \frac{h^3}{\Delta t}\sum_{\ell=1}^{s} h|\psi_{xx}(\tau_{1\ell})|^2 w_\ell^* \Delta t$$

(6. 40)

$$+ \frac{\Delta t}{h}\int_0^{\Delta t} h|\psi_t(\tau)|^2 d\tau\}$$

for $\psi \in P_r([0,h]) \otimes P_s([0,\Delta t])$. Thus, if (6. 35) is valid,

$$\|\nu_x\|_{L^\infty(I\times[0,T])} \le Ch^{-1}G(h,\Delta t) = O(h^{r-1} + (\Delta t)^s),$$

at least. Since

$$\|(u-\hat{u})_x\|_{L^\infty(I\times[0,T])} = O(h^r + (\Delta t)^s),$$

it follows that

(6. 41) $$\|(u-U)_x\|_{L^\infty(I\times[0,T])} \le Ch^{-1}G(h,\Delta t).$$

Consider now the nonlinear equation (1. 1), which can be normalized in the form

(6. 42) $$u_t - a(x,t,u)u_{xx} - b(x,t,u,u_x) = 0, \quad x\in I,\ 0 < t \le T,$$

without loss of generality. The convergence argument will be a perturbation of the one given for the linear equation. First, linearize (6.42) about the solution u to give the linear operator L having coefficients depending on u:

(6.43) $$L = \frac{\partial}{\partial t} - A(x,t)\frac{\partial^2}{\partial x^2} - B(x,t)\frac{\partial}{\partial x} - C(x,t),$$

where

$$A(x,t) = a(x,t,u(x,t)) ,$$

(6.44) $$B(x,t) = b_4(x,t,u(x,t),u_x(x,t)),$$

$$C(x,t) = a_3(x,t,u(x,t))u_{xx}(x,t) + b_3(x,t,u(x,t),u_x(x,t)) ,$$

and, as usual, the subscripts on a and b indicate partial differention. A short calculation shows that, if $\zeta = u - U$,

$$L\zeta = -\frac{1}{2}(a_{33}u_{xx} + b_{33})\zeta^2 - b_{34}\zeta\zeta_x + a_3\zeta\zeta_{xx} - \frac{1}{2}b_{44}\zeta_x^2 ,$$

where the a_{33}, b_{33}, etc., are evaluated where required by the Taylor series remainder. Let $\hat{u}$ be constructed from u and L as in the linear case, using A, B, and C as coefficients. Then,

$$(L\nu)(\xi_{ij},\tau_{k\ell}) = L\{(U-u) + (u-\hat{u})\}(\xi_{ij},\tau_{k\ell})$$

$$= -\rho(\xi_{ij},\tau_{k\ell}) - (q_1\zeta^2 + q_2\zeta\zeta_x + q_3\zeta\zeta_{xx} + q_4\zeta_x^2)(\xi_{ij},\tau_{k\ell}) ,$$

where ρ is calculated for L and the constant in the definition of the $O(\cdot)$ term for ρ given by (6.25) is now of the form

$$C(\|u\|_{2,0}, \|u\|_{1,1})$$

and

$$|q_n(\xi_{ij},\tau_{k\ell})| \leq C(\|u\|_{2,0}) , \quad n = 1,\ldots,4 .$$

The inequality (6.26) can be replaced by

$$-\langle \nu_{xx}, \nu \rangle \Big|_{t_{k-1}}^{t_k} + \frac{m}{2} \sum_{\ell=1}^{s} |\nu_{xx}(\tau_{k\ell})|^2 w_\ell^* \Delta t + \gamma \int_{J_k} |\nu_t(\tau)|^2 d\tau$$

$$\leq C \sum_{\ell=1}^{s} \{ |\rho(\tau_{k\ell})|^2 + |\nu_x(\tau_{k\ell})|^2 + |\nu(\tau_{k\ell})|^2 \} w_\ell^* \Delta t$$

(6.45)

$$+ 2 \sum_{\ell=1}^{s} \{ \langle q_1 \zeta^2, \nu_{xx} \rangle + \langle q_2 \zeta \zeta_x, \nu_{xx} \rangle$$

$$+ \langle q_3 \zeta \zeta_{xx}, \nu_{xx} \rangle + \langle q_4 \zeta_x^2, \nu_{xx} \rangle \} (\tau_{k\ell}) w_\ell^* \Delta t \ .$$

Now, if $\eta = u - \hat{u}$,

$$\zeta^2 = (\eta - \nu)^2 = O(h^{2r+2} + h^{r+1} \nu) + \nu^2$$

and

$$2 |\langle q_1 \zeta^2, \nu_{xx} \rangle| \leq \frac{m}{16} |\nu_{xx}|^2 + Q(\|\nu\|_{0,0}^2 + h^{2r+2}) |\nu|^2 + Qh^{4r+4} \ .$$

Similarly, $\zeta \zeta_x = \nu \nu_x + O(h^{2r+1} + h^r \nu + h^{r+1} \nu_x)$ and

$$2 |\langle q_2 \zeta \zeta_x, \nu_{xx} \rangle| \leq \frac{m}{16} |\nu_{xx}|^2 + Q[(\|\nu\|_{0,0}^2 + h^{2r+2}) |\nu_x|^2 + h^{2r} |\nu|^2 + h^{4r+2}].$$

Next, $\zeta \zeta_{xx} = \nu \nu_{xx} + O(h^{2r} + h^{r-1} \nu + h^{r+1} \nu_{xx})$ and

$$2 |\langle q_3 \zeta \zeta_{xx}, \nu_{xx} \rangle| \leq [Q(\|\nu\|_{0,0} + h^{r+1}) + \frac{m}{16}] |\nu_{xx}|^2 + Q[h^{2r-2} |\nu|^2 + h^{4r}] \ .$$

Finally, $\zeta_x^2 = \nu_x^2 + O(h^{2r} + h^r \nu_x)$ and

$$2 |\langle q_4 \zeta_x^2, \nu_{xx} \rangle| \leq \frac{m}{16} |\nu_{xx}|^2 + Q[(\|\nu_x\|_{0,0}^2 + h^{2r}) |\nu_x|^2 + h^{4r}] \ .$$

Note that no norm of u appears in the $O(\cdot)$ terms above that is not covered by those appearing in the definition of $G(h, \Delta t)$. The above inequalities can be utilized to see that

$$-\langle \nu_{xx}, \nu \rangle \Big|_{t_{k-1}}^{t_k} + \left(\frac{m}{4} - Q\|\nu\|_{0,0}\right) \sum_{\ell=1}^{s} |\nu_{xx}(\tau_{k\ell})|^2 w_\ell^* \Delta t + \gamma \int_{J_k} |\nu_t(\tau)|^2 d\tau$$

(6.46)
$$\leq C\{1 + \|\nu\|_{0,0}^2 + \|\nu_x\|_{0,0}^2\} \sum_{\ell=1}^{s} \{|\rho(\tau_{k\ell})|^2 + |\nu_x(\tau_{k\ell})|^2 + |\nu(\tau_{k\ell})|^2\} w_\ell^* \Delta t$$

$$+ Ch^{4r}\Delta t ,$$

for h sufficiently small. The uniform norms of ν and ν_x are over the set $I \times [0, t_k]$ in (6.46).

In order to complete the convergence analysis, it is necessary only that

$$\max_{\substack{x \in I \\ 0 \leq t \leq t_k}} |\nu(x,t)| \leq m/Q , \qquad \max_{\substack{x \in I \\ 0 \leq t \leq t_k}} |\nu_x(x,t)| \leq 1 ,$$

at least for h and Δt sufficiently small. Let $W = \mathcal{J} u$ and $\psi = W - U$. Then,

(6.47)
$$\{W_t - a(W)W_{xx} - b(W, W_x)\}(\xi_{ij}, \tau_{k\ell}) = \mu(\xi_{ij}, \tau_{k\ell}) ,$$

where

$$\mu(\xi_{ij}, \tau_{k\ell}) = (W-u)_t - a(W)(W-u)_{xx} + [a(u) - a(W)]u_{xx}$$
$$- b_3 \cdot (W-u) - b_4 \cdot (W-u)_x$$

(6.48)
$$= O([1 + \|u\|_{2,0}]\|u\|_{r+2,0} h^r + \|u\|_{r+1,1} h^{r+1}$$
$$+ \|u\|_{r-1,2} h^{r-1} \Delta t + [\|u\|_{0,s+2} + \|u\|_{2,s+1}](\Delta t)^{s+1}),$$

by the estimates leading to (6.18). Now,

$$\psi_t - a(U)\psi_{xx} - (a_3 W_{xx} + b_3)\psi - b_4\psi_x = \mu, \quad (x,t) = (\xi_{ij}, \tau_{k\ell}), \tag{6.49}$$

and $\psi(x,0) = \psi(0,t) = \psi(1,t) = 0$. Since $W_{xx}(\xi_{ij}, \tau_{k\ell}) = O(\|u\|_{2,0})$, it follows from the argument in the linear case that, providing (6.35) is valid,

$$\begin{aligned} \|\psi\|_{L^\infty(I\times[0,T])} &+ h\|\psi_x\|_{L^\infty(I\times[0,T])} \\ &\le C\{\sum_{k=1}^{N}\sum_{\ell=1}^{s} |\mu(\tau_{k\ell})|^2 w_\ell^* \Delta t\}^{1/2} \\ &= O(h^r + h^{r-1}\Delta t + (\Delta t)^{s+1}), \end{aligned} \tag{6.50}$$

with the constant depending on the norms of u that appear in the definition of ρ. Since

$$\nu = (W-\hat{u}) + (U-W) = h^{r+2}D - \psi,$$

it follows that the crude estimate

$$\|\nu\|_{L^\infty(I\times[0,T])} + h\|\nu_x\|_{L^\infty(I\times[0,T])} = O(h^r + h^{r-1}\Delta t + (\Delta t)^{s+1}) \tag{6.51}$$

holds. It then follows from (6.46), (6.51), and the argument used in the linear case that

$$\begin{aligned} \|\nu\|_{L^\infty(\times[0,T])} &+ h\|\nu_x\|_{L^\infty(I\times[0,T])} \\ &\le M(\|u\|_{r+3,0}, \|u\|_{r+2,1}, \|u\|_{r-1,2}, \|u\|_{0,s+2}, \|u\|_{2,s+1}) \cdot \\ &\quad \cdot (h^{r+1} + h^{r-1}\Delta t + (\Delta t)^{s+1}) \\ &\equiv G_1(h,\Delta t). \end{aligned} \tag{6.52}$$

Hence, the following theorem has been proved.

Theorem 6.2. Let u be the solution of (6.42) and let $U \in \mathcal{M}_1(r,\delta)\otimes\mathcal{M}_0(s,\epsilon)$ be the solution of the collocation equations (6.16) corresponding to (6.42). Assume that a, b, and c have bounded third derivatives in a neighborhood of the solution u and that the norms of u appearing in the definition (6.52) of $G_1(h,\Delta t)$ are finite. Let (6.35) hold. Then,

$$\|u-U\|_{L^\infty(I\times[0,T])} + h\|(u-U)_x\|_{L^\infty(I\times[0,T])} \le CG_1(h,\Delta t).$$

The algebraic problem at each time step associated with (6.16) is nonlinear if the differential equation is nonlinear. We saw earlier that it was feasible to extrapolate the coefficients in the Crank-Nicolson-collocation method to produce linear algebraic equations without loss of asymptotic order of accuracy. We shall now see that it is also feasible to linearize the computational problem for full collocation by extrapolating the coefficients. Consider first the slightly simplified differential problem

$$(6.53)\qquad \begin{aligned} &u_t - a(x,t,u)u_{xx} - b(x,t,u)u_x - c(x,t,u) = 0, \quad x\in I,\ t\in J=[0,T],\\ &u(0,t) = u(1,t) = 0, \quad t\in J,\\ &u(x,0) = f(x), \quad x\in I. \end{aligned}$$

The important restriction is the appearance of u_x linearly; the vanishing of u on the boundary is for convenience only. Modify the collocation equations to read

$$\{U_t - a(x,t,V)U_{xx} - b(x,t,V)U_x - c(x,t,V)\}(\xi_{ij}, \tau_{k\ell}) = 0, \quad i = 1,\dots,M,$$
$$j = 1,\dots,r-1,$$

(6.54) $$U(0,t) = U(1,t) = 0, \quad t \in J, \qquad k = 2,\dots,N,$$

$$U(x,t) \text{ determined by (6.16)}, \ (x,t) \in I \times J_1, \qquad \ell = 1,\dots,s,$$

where $V \in \mathfrak{M}_1(r,\delta) \otimes P_s(J_{k-1} \cup J_k)$ is the direct extrapolation of U from its values in J_{k-1}:

(6.55) $$V(x,t) = U_{J_{k-1}}(x,t), \quad x \in I, \ t \in J_k.$$

Note that this procedure reduces essentially to (5.12) for the case $s = 1$.

Let

$$A(x,t) = a(x,t,u(x,t)),$$

$$B(x,t) = b(x,t,u(x,t)),$$

$$C(x,t) = a_3(x,t,u(x,t))u_{xx}(x,t) + b_3(x,t,u(x,t))u_x(x,t) + c_3(x,t,u(x,t)),$$

and

$$L = \frac{\partial}{\partial t} - A\frac{\partial^2}{\partial x^2} - B\frac{\partial}{\partial x}.$$

If $\zeta = u-U$ and $\mu = u-V$, then differencing (6.53) and (6.54) leads to the relation

(6.56) $$(L\zeta)(\xi_{ij}, \tau_{k\ell}) = \{C\mu - \frac{1}{2}(a_{33}^* u_{xx} + b_{33}^* u_x + c_{33}^*)\mu^2 - a_3^* \zeta_{xx}\mu - b_3^* \zeta_x \mu\}(\xi_{ij}, \tau_{k\ell}),$$

where the asterisk indicates evaluation as required by the remainder for the Taylor expansion. Now construct $\hat{u}$ from u and the operator L just defined, and set $\nu = U-\hat{u}$ as before. Then,

(6.57) $$(L\nu)(\xi_{ij}, \tau_{k\ell}) = (-\rho + C\mu + q_1\mu^2 + q_2\zeta_{xx}\mu + q_3\zeta_x\mu)(\xi_{ij}, \tau_{k\ell}),$$

where ρ is defined as in (6.25).

Let us estimate μ. First, for $t \in J_{k-1} \cup J_k$,

$$u(x,t) = \sum_{\alpha=0}^{s} \frac{1}{\alpha!} \frac{\partial^\alpha u}{\partial t^\alpha}(x, t_{k-1})(t-t_{k-1})^\alpha + O(\|u\|_{0,s+1}(\Delta t)^{s+1})$$

$$= p_s(x,t) + O(\|u\|_{0,s+1}(\Delta t)^{s+1}), \quad p_s(x,\cdot) \in P_s(J_{k-1} \cup J_k).$$

Thus,

$$\mu(x,t) = [p_s(x,t) - V(x,t)] + O(\|u\|_{0,s+1}(\Delta t)^{s+1}), \quad x \in I,\ t \in J_k.$$

It is clear that there exists a constant $Q = Q(s)$ such that

$$\|w\|_{L^\infty([1,2])} \leq Q\|w\|_{L^\infty([0,1])}, \quad w \in P_s([0,2]);$$

again by homogeneity

$$\|\mu\|_{L^\infty(I\times J_k)} \leq Q[\|p_s - V\|_{L^\infty(I\times J_{k-1})} + \|u\|_{0,s+1}(\Delta t)^{s+1}]$$

$$\leq Q[\|\zeta\|_{L^\infty(I\times J_{k-1})} + \|u\|_{0,s+1}(\Delta t)^{s+1}].$$

Let $\eta = u-\hat{u}$ as usual. Then by (6.38),

$$\|\zeta\|_{L^\infty(I\times J_{k-1})} \leq \|\eta\|_{L^\infty(I\times J_{k-1})} + \|\nu\|_{L^\infty(I\times J_{k-1})}$$

$$\leq \|\nu\|_{L^\infty(I\times J_{k-1})} + Q[(\|u\|_{r+1,0} + h\|u\|_{r+2,0})h^{r+1}$$

$$+ \|u\|_{0,s+1}(\Delta t)^{s+1}]$$

for $r \geq 4$ (and slightly different for $r = 3$). Thus,

$$\|\mu\|_{L^{\infty}(I\times J_k)} \leq Q[\|\nu\|_{L^{\infty}(I\times J_{k-1})} + (\|u\|_{r+1,0} + h\|u\|_{r+2,0})h^{r+1} + \|u\|_{0,s+1}(\Delta t)^{s+1}] . \tag{6.58}$$

Since L^{∞} bounds will be needed for ν, assume that (6.35) holds. The derivation of (6.26) can be repeated starting from (6.57) to give

$$\begin{aligned} -<\nu_{xx},\nu>\Big|_{t_{k-1}}^{t_k} &+ \frac{m}{2}\sum_{\ell=1}^{s}|\nu_{xx}(\tau_{k\ell})|^2 w_\ell^* \Delta t + \gamma\int_{J_k}|\nu_t(\tau)|^2 d\tau \\ &\leq Q\sum_{\ell=1}^{s}\{|\rho|^2 + |\nu|^2 + |\nu_x|^2\}(\tau_{k\ell})w_\ell^*\Delta t \\ &+ 2\sum_{\ell=1}^{s}\{<C\mu,\nu_{xx}> + <q_1\mu^2 + q_2\zeta_{xx}\mu + q_2\zeta_x\mu,\nu_{xx}>\}(\tau_{k\ell})w_\ell^*\Delta t . \end{aligned} \tag{6.59}$$

Now, with the constants in the $O(\cdot)$-terms being indicated from (6.58) and $|\eta_x|$ and $|\eta_{xx}|$ being dominated by $O(h^r + (\Delta t)^{s+1})$,

$$|2<C\mu,\nu_{xx}>| \leq \frac{m}{32}|\nu_{xx}|^2 + Q[\|\nu\|^2_{L^{\infty}(I\times J_{k-1})} + O(h^{2r+2} + (\Delta t)^{2s+2})],$$

$$|2<q_1\mu^2,\nu_{xx}>| \leq \frac{m}{32}|\nu_{xx}|^2 + Q[\|\nu\|^4_{L^{\infty}(I\times J_{k-1})} + O(h^{4r+4} + (\Delta t)^{4s+4})],$$

$$|2<q_2\zeta_{xx}\mu,\nu_{xx}>| \leq \frac{m}{32}|\nu_{xx}|^2 + Q[\|\nu\|^2_{L^{\infty}(I\times J_{k-1})} + O(h^{2r+2} + (\Delta t)^{2s+2})] \cdot [|\nu_{xx}|^2 + O(h^{2r} + (\Delta t)^{2s+2})],$$

$$|2<q_3\zeta_x\mu,\nu_{xx}>| \leq \frac{m}{32}|\nu_{xx}|^2 + Q[\|\nu\|^2_{L^{\infty}(I\times J_{k-1})} + O(h^{2r+2} + (\Delta t)^{2s+2})] \cdot [|\nu_x|^2 + O(h^{2r} + (\Delta t)^{2s+2}] .$$

Thus, for h and Δt sufficiently small and for $k = 2, \dots, N$,

$$- \langle \nu_{xx}, \nu \rangle \Big|_{t_{k-1}}^{t_k} + \Big(\frac{m}{4} - Q\|\nu\|^2_{L^\infty(I\times J_{k-1})}\Big) \sum_{\ell=1}^{s} |\nu_{xx}(\tau_{k\ell})|^2 w_\ell^* \Delta t + \gamma \int_{J_k} |\nu_t(\tau)|^2 d\tau$$

(6.60)

$$\leq Q(1 + \|\nu\|^2_{L^\infty(I\times J_{k-1})}) \Big[\sum_{\ell=1}^{s} \{|\rho|^2 + |\nu|^2 + |\nu_x|^2\}(\tau_{k\ell}) w_\ell^* \Delta t + \|\nu\|^2_{L^\infty(I\times J_{k-1})} \Delta t$$

$$+ Q[(\|u\|_{r+1,0} + h\|u\|_{r+2,0})^2 h^{2r+2} + \|u\|^2_{0,s+1} (\Delta t)^{2s+2}] \Delta t .$$

For $k = 1$, (6.46) holds with the uniform norms of ν and ν_x being taken over $I \times J_1$. The previous argument showed that

$$\|\nu\|_{L^\infty(I\times J_1)} \leq QG_1(h, \Delta t) = O(h^{r+1} + h^{r-1}\Delta t + (\Delta t)^{s+1});$$

see (6.52). The argument used in the linear case can be combined with induction on k for $2 \leq k \leq N$ to show that again

(6.61) $$\|\nu\|_{L^\infty(I\times J)} + h\|\nu_x\|_{L^\infty(I\times J)} \leq QG_1(h, \Delta t),$$

and we have proved the following theorem.

Theorem 6.3. Let u be the solution of (6.53) and let $U \in \mathcal{M}_1(r,\delta)\otimes\mathcal{M}_0(s,\epsilon)$ be the solution of the extrapolated collocation equations (6.54)-(6.55). If a, b, and c have bounded third derivatives in the neighborhood of u, if the norms of u appearing in the definition (6.52) of $G_1(h,\Delta t)$ are finite, and if (6.35) holds, then

$$\|u-U\|_{L^\infty(I\times J)} + h\|(u-U)_x\|_{L^\infty(I\times J)} \le CG_1(h,\Delta t).$$

Since U is found on $I\times J_1$ by using the nonlinear equations (6.16), it is necessary to use an iterative procedure to get started. The same code that is used for solving (6.54) can be employed with a successive substitution for evaluating the coefficients ; this iteration converges for sufficiently small Δt, independently of h. More complicated procedures could also be used.

The more general differential equation (6.42) can be treated by a linearized collocation method by replacing (6.54) by

$$(6.62)\quad \{U_t - a(V)U_{xx} - b(V,V_x) - b_4(V,V_x)(U_x - V_x)\}(\xi_{ij},\tau_{k\ell}) = 0,\quad \begin{aligned} i &= 1,\dots,M,\\ j &= 1,\dots,r-1,\\ k &= 2,\dots,N,\\ \ell &= 1,\dots,s, \end{aligned}$$

where V is defined as before by (6.55) and U is determined on $I\times J_1$ by (6.16). An analysis of the error caused by (6.62) can be made in a similar manner to that given for (6.54), and the analogue of Theorem 6.3 will result.

7. Some Extensions. A few slightly more general methods should be mentioned briefly. First, we have throughout used piecewise-polynomial functions with the polynomials having the same degree in each subinterval I_i. Clearly, this is not necessary. Let

$$(7.1)\qquad \mathfrak{M}_1(r_1, \ldots, r_N, \delta) = \{v \in C^1(I) \mid v \in P_{r_i}(I_i),\ i = 1, \ldots, M\} .$$

Let $\xi_1^{(r)}, \ldots, \xi_{r-1}^{(r)}$ denote the roots of the Legendre polynomial of degree $r-1$ on I, and set

$$(7.2)\qquad \xi_{ij} = x_{i-1} + h_i \xi_j^{(r_i)},\quad j = 1, \ldots, r_i - 1,\quad i = 1, \ldots, M.$$

Then generalize (4.1) by requiring that

$$(7.3)\qquad \{c(U)\frac{\partial U}{\partial t} - a(U)\frac{\partial^2 U}{\partial x^2} - b(U, U_x)\}(\xi_{ij}, t) = 0,\quad j = 1, \ldots, r_i - 1,\quad i = 1, \ldots, M .$$

The entire analysis of method (4.1) was essentially done on subintervals. It is obvious that the definition of the interpolation operator $S_{r,\delta}$ extends to the present case and that Lemma 2.3 remains valid. The convergence proof leading to Theorem 4.1 can be repeated to show that

$$(7.4)\qquad \|u-U\|_{L^\infty(I\times[0,T])} \le C\{\sum_{i=1}^{M} (\|u\|^2_{L^\infty(0,T;H^{r_i+3}(I_i))} + \|u_t\|_{L^2(0,T;H^{r_i+3}(I_i))})h_i^{2r_i+2}\}^{1/2},$$

assuming that $U(x, 0)$ is obtained from the obvious generalization of $T_{r,\delta}$ applied to f. The discretizations in time will be left to the reader.

A more interesting example occurs when the coefficients of the differential equation are discontinuous along certain lines $x = \text{constant}$. It suffices to describe the case of a single discontinuity, say at $x = X \in (0, 1)$. Assume the differential equation in the form

$$(7.5) \qquad c(x)\frac{\partial u}{\partial t} - \frac{\partial}{\partial x}\left(a(x)\frac{\partial u}{\partial x}\right) = 0 ,$$

where c and a are smooth in $\{x < X\}$ and $\{x > X\}$ and have one-sided limits from either side at $x = X$. Then under reasonable assumptions on the remaining data of the problem the solution is smooth in both $\{x<X\}$ and $\{x>X\}$ and is continuous on the whole region, along with au_x. Now, assume that δ is so chosen that $X = x_i$. The usual choice of a basis for the space $\mathcal{M}_1(r, \delta)$ uses two knot-oriented functions (i.e., the value and slope basis functions for $\mathcal{M}_1(3, \delta)$) at each knot and $r-3$ interval-oriented functions (i.e., $(x-x_{k-1})^2(x-x_k)^2(x-x_{k-1/2})^\ell$, $\ell = 0, \ldots, r-4$) in each interval. If the slope basis element W at $x_i = X$ is modified so that $a(x_i-0)W'(x_i-0) = a(x_i+0)W'(x_i+0)$ and all other basis elements are left unmodified, then the modified space, say $\tilde{\mathcal{M}}_1(r, \delta)$, can be used in place of $\mathcal{M}_1(r, \delta)$ in the formulation of the collocation method. It is clear that $T_{r,\delta}u$, as pieced together from its local definition, lies in $\tilde{\mathcal{M}}(r, \delta)$. Moreover,

$$(7.6) \qquad \|u-T_{r,\delta}u\|_{L^\infty(I\times[0,T])} \le C\left\{\sum_{k=1}^{M} \|u\|^2_{L^\infty(0,T;H^{r+2}(I_k))} h_k^{2r+2}\right\}^{1/2} .$$

The estimate (4.13) remains valid in this case; thus, optimal order convergence is obtainable for piecewise-smooth solutions. Again, it is feasible to use varying degrees for the polynomials in the subintervals. Time-discretization is unaffected by the change to $\tilde{\mathfrak{M}}_1(r, \delta)$.

CHAPTER II

SUPERCONVERGENCE ESTIMATES AT THE KNOTS

8. Introduction. Since the global estimates of Chapter I were all optimal order estimates, no faster rates of convergence can be expected to hold in the entire domain $I \times [0, T]$. However, it is possible that distinguishable points can be located at which the convergence rate is faster than the global rate. DeBoor and Swartz [6] have shown that the error at the knots $x = x_i$ for the collocation method for the two-point boundary-value problem using $\mathcal{M}_1(r, \delta)$ and the Gauss points ξ_{ij} is $O(h^{2r-2})$ in the uniform interval case. The object of this chapter is to extend this result to the parabolic problem. A special case was presented in [7].

Section 9 studies the two-point boundary-value problem. The DeBoor-Swartz results were reproven by a new method of proof based on the construction of a quasi-interpolant associated with the differential operator. Both the function and the slope are approximated to within $O(h^{2r-2})$ at the knots. The quasi-interpolant approach is then applied in section 10 to the continuous-time collocation (4.1). The simple choice of the initial condition given by $T_{r,\delta}f$ is not adequate to obtain the superconvergence on the lines $x = x_i$, but a more complicated initial condition based onthe quasi-interpolant can be constructed practically and can be used. We show that the error in the values of the approximation is $O(h^{2r-2})$ on the knot lines in the nonlinear case; a corollary of this result is that the error in the slope on the knot lines is $O(h^{2r-5/2})$. We also obtain the improved estimate $O(h^{2r-2})$ for the knot line slope errors in the case that (1.1) is linear. Undoubtedly, the improved estimate also holds for the nonlinear problem, but our energy flagged. In section 11 the quasi-interpolant is refined to give us knot error error estimates of the form $O(h^{2r-2}+(\Delta t)^{2s})$ for the full collocation method (6.16); the analysis is given only for the linear parabolic equation. The construction of the quasi-interpolant is a tedious process; elsewhere it will be shown [13,14] that this argument can be applied to obtain superconvergence at the knots for the Galerkin method for (1.1) based on using the space $\mathcal{M}_0(r,\delta)$ in space.

The results of this chapter are the foundation of the improved local estimates of the next chapter in which the space $\mathcal{M}_1(r,\delta)$ will be modified locally so as to allow better uniform approximation on subdomains of I. They also lead to the eigenvalue estimates of Chapter IV. The authors also believe that this chapter is the most interesting one of the manuscript.

There are a number of related superconvergence results for Galerkin methods; see [2, 3, 10-15, 20, 21].

A brief appendix ends this chapter; in it is an indication of how to construct the function $\hat{u}$ used in section 6, (6.19)-(6.25), in the case $r = 3$.

9. A Two-Point Boundary Problem. Consider the two-point boundary problem

$$(9.1)\qquad \begin{aligned} &Ly = y'' + a(x)y' + b(x)y = f(x), \quad x \in I = [0,1], \\ &y(0) = y(1) = 0, \end{aligned}$$

and the associated collocation method of finding $Y \in \mathcal{M}_1(r,\delta)$ such that

$$(9.2)\qquad \begin{aligned} &LY(\xi_{ij}) = (Y'' + aY' + bY)(\xi_{ij}) = f(\xi_{ij}), \quad i = 1,\dots,M,\ j = 1,\dots,r-1, \\ &Y(0) = Y(1) = 0. \end{aligned}$$

For convenience assume that the coefficients $a(x)$ and $b(x)$ are smooth and that

$$(9.3)\qquad \|v\|_{H^2(I)} \leq C\|Lv\|_{L^2(I)}, \quad v \in H^2(I) \cap H_0^1(I),$$

and that δ is uniform; i.e., $h_i = h$. Existence and uniqueness of Y follow from (9.3) and Lemma 9.3 below for sufficiently small h. Let $T_{r,\delta}: C^1(I) \to \mathcal{M}_1(r,\delta)$ be the interpolation operator defined in §2. Recall that for $x \in I_i$

$$(9.4)\qquad (y - T_{r,\delta}y)(x) = \sum_{q=1}^{p} y^{(r+q)}(x_{i-1/2})C_q\left(\frac{x-x_{i-1}}{h}\right)h^{r+q} + S_{r,p,\delta}(x),$$

where

$$(9.5)\qquad \|S^{(\alpha)}_{r,p,\delta}\|_{L^\infty(I_i)} \leq C\|y^{(r+p+1)}\|_{L^\infty(I_i)} h^{r+p+1-\alpha}, \quad 0 \leq \alpha \leq r+p.$$

It then is an easy calculation to see that, since $C_1''(\xi_j) = 0$,

$$
\begin{aligned}
L(y-T_{r\delta}y)(\xi_{ij}) = & \sum_{q=2}^{r-1} y^{(r+q)}(x_{i-1/2})C_q''(\xi_j)h^{r+q-2} \\
& + a(\xi_{ij}) \sum_{q=1}^{r-2} y^{(r+q)}(x_{i-1/2})C_q'(\xi_j)h^{r+q-1} \\
& + b(\xi_{ij}) \sum_{q=1}^{r-3} y^{(r+q)}(x_{i-1/2})C_q(\xi_j)h^{r+q} \\
& + O(\|y\|_{W^{2r}(I_i)} h^{2r-2}) ,
\end{aligned} \tag{9.6}
$$

where

$$
\|y\|_{W^k(E)} = \sum_{\ell=0}^{k} \|y^{(\ell)}\|_{L^\infty(E)} . \tag{9.7}
$$

Recall the orthogonalities (2.9) and also expand $a(x)$ and $b(x)$ about $x_{i-1/2}$. Then, $L(y - T_{r,\delta}y)$ can be expressed in the form

$$
L(y-T_{r,\delta}y)(\xi_{ij}) = \sum_{q=0}^{r-3} h^{r+q}R_{iq}^{(0)}(\xi_j) + E_i^{(0)}(\xi_j)h^{2r-2} , \tag{9.8}
$$

where

$$
\begin{aligned}
|R_{iq}^{(0)}(\xi_j)| & \leq C\|y\|_{W^{r+q+2}(I_i)} , \\
|E_i^{(0)}(\xi_j)| & \leq C\|y\|_{W^{2r}(I_i)} , \\
{}_1\langle R_{iq}^{(0)}, x^k\rangle & = 0 , \quad 0 \leq k \leq r-q-3 ,
\end{aligned} \tag{9.9}
$$

since for $x \in I$ and $q = 0, \ldots, r-3$,

$$R_{iq}^{(0)}(x) = y^{(r+q+2)}(x_{i-1/2})C''_{q+2}(x)$$

(9.10)
$$+ \sum_{\ell=0}^{q} \frac{1}{\ell!} a^{(\ell)}(x_{i-1/2}) y^{(r+q+1-\ell)}(x_{i-1/2})(x-1/2)^{\ell} C'_{q+1-\ell}(x)$$

$$+ \sum_{\ell=0}^{q-1} \frac{1}{\ell!} b^{(\ell)}(x_{i-1/2}) y^{(r+q-\ell)}(x_{i-1/2})(x-1/2)^{\ell} C_{q-\ell}(x) \quad ;$$

the sum of terms involving b is missing for $q = 0$.

It proved helpful in the analysis of the full collocation procedure (6.16) to construct a quasi-interpolant $\hat{u}$ of the solution u of the differential equation so that the differential operator applied to $u-\hat{u}$ produced a small residual at the collocation points. This approach will be carried out to a greater depth here. First, let us prove two generalizations of Lemmas 6.1 and 6.2.

<u>Lemma 9.1.</u> Let $\langle F, 1 \rangle_i = 0$, $i = 1, \ldots, M$. There exists a unique $D \in \mathcal{M}_1(r, \delta)$ such that

$$h^2 D''(\xi_{ij}) = F(\xi_{ij}), \quad i = 1, \ldots, M, \quad j = 1, \ldots, r-1,$$

$$D(0) = D(1) = 0 \; ;$$

moreover,

$$h\|D\|_{W^1(I)} + h^2 \|D''\|_{L^\infty(I)} \le C \|F\|_{L^\infty(I)} .$$

<u>Proof.</u> Let $d_i \in \mathcal{M}_1(r, \delta)$ be constructed by requiring that

i) $d_i(x) = 0 \quad , \quad x \le x_{i-1}$,

ii) $h^2 d_i''(\xi_{ij}) = F(\xi_{ij}) \quad , \quad j = 1, \ldots, r-1$,

iii) $d_i(x_{i-1}) = d_i'(x_{i-1}) = d_i'(x_i) = 0$,

iv) $d_i(x) = d_i(x_i) \quad , \quad x \ge x_i$.

Note that $d_i'' \in P_{r-2}(I_i)$ is uniquely determined by ii). Then

$$d_i(x) = \int_{x_{i-1}}^{x} (x-t) d_i''(t) dt \quad , \quad x \in I_i \; ,$$

satisfies the first two requirements of iii), and the assumption $\langle F, 1 \rangle_i = 0$ implies that

$$d_i'(x_i) = \int_{x_{i-1}}^{x_i} d_i''(t) dt = \langle F, 1 \rangle_i = 0 \; .$$

It is clear that $d_i \in \mathcal{M}_1(r, \delta)$.

Set

$$D(x) = \sum_{i=1}^{M} \{ d_i(x) - x d_i(1) \} \; .$$

Then $D \in \mathcal{M}_1(r, \delta)$, $D(0) = D(1) = 0$, and $h^2 D''(\xi_{ij}) = h^2 d_i''(\xi_{ij}) = F(\xi_{ij})$. This establishes the existence of the desired function. Homogeneity in h shows that

$$|h^k d_i^{(k)}(x)| \le C \|F\|_{L^\infty(I_i)} \quad , \quad k = 0, 1.$$

Thus, the a priori bound for D in $W^1(I)$ follows readily, and uniqueness is an obvious corollary of the bound.

Lemma 9.2. Let $r \geq 4$ and $1 \leq q \leq r-3$. Let $<F, x^k>_i = 0$, $i = 1, \ldots, M$, $k = 0, \ldots, q$. Then there exists a unique $D \in \mathcal{M}_1(r, \delta)$ such that

$$h^2 D''(\xi_{ij}) = F(\xi_{ij}) , \quad i = 1, \ldots, M, \quad j = 1, \ldots, r-1,$$

$$D(x_i) = D'(x_i) = 0 \quad , \quad i = 0, \ldots, M ,$$

and

$$\|D\|_{L^\infty(I)} + h\|D'\|_{L^\infty(I)} + h^2\|D''\|_{L^\infty(I)} \leq C\|F\|_{L^\infty(I)} .$$

Also,

$$<D', x^k>_i = 0, \quad i = 1, \ldots, M, \quad k = 0, \ldots, q-1 .$$

If $q \geq 2$,

$$<D, x^k>_i = 0, \quad i = 1, \ldots, M, \quad k = 0, \ldots, q-2 .$$

Proof. Let d_i and D be the same as in the proof of Lemma 9.1. Note that

$$d_i(x_i) = \int_{x_{i-1}}^{x_i} (x_i - t) d_i''(t) dt = <F(t), x_i - t>_i = 0 .$$

Since the supports of the d_i's do not overlap, the a priori estimate holds. Observe that $D'(x)x^k \in P_{r+k-1}(I_i) \subset P_{2r-3}(I_i)$ for $k \leq r-2$. Thus,

$$<D', x^k>_i = \int_{I_i} D'(x)x^k dx = -\frac{1}{k+1}\int_{I_i} D''(x)x^{k+1} dx$$

$$= -\frac{1}{k+1} <F, x^{k+1}>_i = 0 , \quad k \leq q-1 .$$

Similarly, if $q \geq 2$ and $k \leq r-3$, $D(x)x^k \in P_{2r-3}(I_i)$ and

$$< D, x^k >_i = \frac{1}{(k+1)(k+2)} < F, x^{k+2} >_i = 0 \ , \quad k \leq q-2,$$

and the lemma is demonstrated.

The construction of the quasi-interpolant can begin. Let D_0 be the unique function in $\mathcal{M}_1(r, \delta)$ such that

(9.11a) $$h^2 D''(\xi_{ij}) = R_{i0}^{(0)}(\xi_j) \ , \quad i = 1, \ldots, M, \ j = 1, \ldots, r-1,$$

and either

(9.11b) $$D_0(x_i) = D_0'(x_i) = 0 \ , \quad i = 0, \ldots, M \ \text{ for } \ r \geq 4,$$

or

(9.11c) $$D_0(0) = D_0(1) = 0 \quad , \ \text{for} \ \ r = 3.$$

The existence of D_0 is guaranteed by (9.9) and either Lemma 9.1 or Lemma 9.2. Also,

(9.12a) $$\|D_0\|_{L^\infty(I)} + h\|D_0'\|_{L^\infty(I)} \leq C\|y\|_{W^{r+2}(I)} \ , \quad r \geq 4 \ ,$$

or

(9.12b) $$h\|D_0\|_{L^\infty(I)} + h\|D_0'\|_{L^\infty(I)} \leq C\|y\|_{W^5(I)} \ , \quad r = 3.$$

Let $\tilde{y}^{(0)} = T_{r,\delta} y + h^{r+2} D_0$. Then, for $r = 3$,

(9.13) $$\begin{aligned} L(y-\tilde{y}^{(0)})(\xi_{ij}) &= E_i^{(0)}(\xi_j)h^4 - a(\xi_{ij})D_0'(\xi_{ij})h^5 - b(\xi_{ij})D_0(\xi_{ij})h^5 \\ &= E_i^{(1)}(\xi_j)h^4 \ , \end{aligned}$$

where

(9.14) $$|E_i^{(1)}(\xi_j)| \leq C\|y\|_{W^6(I)} \ .$$

For $r \geq 4$,

$$L(y-\tilde{y}^{(0)})(\xi_{ij}) = \sum_{q=1}^{r-3} R_{iq}^{(1)}(\xi_j)h^{r+q} + E_i^{(1)}(\xi_j)h^{2r-2} , \tag{9.15}$$

where

$$\begin{aligned} R_{iq}^{(1)}(\xi_j) = R_{iq}^{(0)}(\xi_j) &- \frac{1}{(q-1)!} a^{(q-1)}(x_{i-1/2})hD_0'(\xi_{ij})\left(\frac{\xi_{ij} - x_{i-1/2}}{h}\right)^{q-1} \\ &- \frac{1}{(q-2)!} b^{(q-2)}(x_{i-1/2})D_0(\xi_{ij})\left(\frac{\xi_{ij} - x_{i-1/2}}{h}\right)^{q-2} ; \end{aligned} \tag{9.16}$$

derivatives of negative order are to be interpreted as zero. It is easy to see from Lemma 9.2 that

$$\begin{aligned} {}_1\langle R_{iq}^{(1)}, x^k \rangle &= 0 , \quad 0 \leq k \leq r-q-3 , \\ |R_{iq}^{(1)}(\xi_j)| &\leq C\|y\|_{W^{r+q+2}(I)} , \\ |E_i^{(1)}(\xi_j)| &\leq C\|y\|_{W^{2r}(I)} ; \end{aligned} \tag{9.17}$$

i.e., the constraints on $R_{iq}^{(1)}$ and $E_i^{(1)}$ are of exactly the same form as those on $R_{iq}^{(0)}$ and $E_i^{(0)}$.

If $r > 4$, the above construction can be iterated for $\nu = 1, \ldots, r-4$ as follows. Let $D_\nu \in \mathcal{M}_1(r, \delta)$ be found from the equations

$$\begin{aligned} h^2 D_\nu''(\xi_{ij}) &= R_{i\nu}^{(\nu)}(\xi_j), \quad i = 1, \ldots, M, \; j = 1, \ldots, r-1, \\ D_\nu(x_i) &= D_\nu'(x_i) = 0 , \quad i = 0, \ldots, M . \end{aligned} \tag{9.18}$$

Set

$$\tilde{y}^{(\nu)} = T_{r,\delta}y + \sum_{k=0}^{\nu} h^{r+k+2}D_k . \tag{9.19}$$

If, for $q = \nu+1, \ldots, r-3$,

$$R_{iq}^{(\nu+1)}(\xi_j) = R_{iq}^{(\nu)}(\xi_j) - \frac{1}{(q-\nu-1)!} a^{(q-\nu-1)}(x_{i-1/2}) h D_\nu'(\xi_{ij}) \left(\frac{\xi_{ij} - x_{i-1/2}}{h}\right)^{q-\nu-1}$$

(9.20)

$$- \frac{1}{(q-\nu-2)!} b^{(q-\nu-2)}(x_{i-1/2}) D_\nu(\xi_{ij}) \left(\frac{\xi_{ij} - x_{i-1/2}}{h}\right)^{q-\nu-2},$$

then

(9.21) $$L(y-\tilde{y}^{(\nu)})(\xi_{ij}) = \sum_{q=\nu+1}^{r-3} R_{iq}^{(\nu+1)}(\xi_j) h^{r+q} + E_i^{(\nu+1)}(\xi_j) h^{2r-2} .$$

It again follows from Lemma 9.2 that

$$_1\langle R_{iq}^{(\nu+1)}, x^k \rangle = 0, \quad 0 \le k \le r-q-3,$$

(9.22) $$|R_{iq}^{(\nu+1)}(\xi_j)| \le C \|y\|_{W^{r+q+2}(I)} ,$$

$$|E_i^{(\nu+1)}(\xi_j)| \le C \|y\|_{W^{2r}(I)} .$$

Finally, use Lemma 9.1 to construct the non-local correction $D_{r-3} \in \mathcal{M}_1(r, \delta)$ such that

$$h^2 D_{r-3}''(\xi_{ij}) = R_{i,r-3}^{(r-3)}(\xi_j) , \quad i = 1, \ldots, M, \quad j = 1, \ldots, r-1 ,$$

(9.23)

$$D_{r-3}(0) = D_{r-3}(1) = 0 ,$$

and set

(9.24) $$\hat{y} = \tilde{y}^{(r-3)} = T_{r,\delta} y + \sum_{k=0}^{r-3} h^{r+k+2} D_k .$$

Then,

(9.25) $$L(y-\hat{y})(\xi_{ij}) = E_i^{(r-2)}(\xi_j) h^{2r-2} = \rho(\xi_{ij}) ,$$

where

$$\rho(\xi_{ij}) = E_i^{(r-3)}(\xi_j)h^{2r-2} - a(\xi_{ij})h^{2r-1}D'_{r-3}(\xi_{ij}) - b(\xi_{ij})h^{2r-1}D_{r-3}(\xi_{ij})$$

(9.26)

$$= O(\|y\|_{W^{2r}(I)} h^{2r-2}) .$$

Before proceeding, it should be noted that the case $r = 3$ satisfies (9.24)-(9.26); see (9.13) and (9.14).

The function $\hat{y} \in \mathcal{M}_1(r, \delta)$ is the desired quasi-interpolant of y. We wish now to show that the collocation solution Y is very nearly $\hat{y}$. Since $L(y-Y)(\xi_{ij}) = 0$ from the definition of the collocation method (9.2),

$$L(Y-\hat{y})(\xi_{ij}) = \rho(\xi_{ij}), \quad i = 1, \ldots, M, \; j = 1, \ldots, r-1,$$

(9.27)

$$(Y-\hat{y})(0) = (Y-\hat{y})(1) = 0.$$

The next lemma will allow us to find a bound for $Y-\hat{y}$ in $H^2(I)$.

Lemma 9.3. There exists a constant C such that

$$|< Lw, Lw > - \|Lw\|^2_{L^2(I)}| \leq Ch\|w\|^2_{H^2(I)}$$

for all $w \in \mathcal{M}_1(r, \delta)$.

Proof. The proof will result from a term-by-term consideration of the expansion

$$< Lw, Lw > = |w''|^2 + |aw'|^2 + |bw|^2 + 2< w'', aw' > + 2< w'', bw >$$

$$+ 2< aw', bw > .$$

Since $w'' \in P_{r-2}(I_i)$,

$$|w''|^2 = \|w''\|^2_{L^2(I)} .$$

Let $\bar{a}_i = h^{-1}\langle a, 1\rangle_i$, $\bar{b}_i = h^{-1}\langle b, 1\rangle_i$, etc. Then,

$$\langle aw'', w'\rangle = \sum_{i=1}^{M} \langle aw'', w'\rangle_i$$

$$= \sum_{i=1}^{M} \bar{a}_i \int_{I_i} w''w'\,dx + \sum_{i=1}^{M} \langle (a - \bar{a}_i)w'', w'\rangle_i$$

$$= \int_I aw''w'\,dx + \sum_{i=1}^{M} \{\langle (a - \bar{a}_i)w'', w'\rangle_i - \int_{I_i} (a - \bar{a}_i)w''w'dx\}$$

$$= (aw'', w') + O(h \sum_{i=1}^{M} \|w''\|_{L^2(I_i)} \|w'\|_{L^2(I_i)})$$

$$= (aw'', w') + O(h\{\|w''\|^2_{L^2(I)} + \|w'\|^2_{L^2(I)}\}) ,$$

using Lemma 3.3 and the smoothness of $a(x)$.

Next, look at

$$\langle bw'', w\rangle = \sum_{i=1}^{M} \{\langle \bar{b}_i w'', w\rangle_i + O(h\|w''\|_{L^2(I_i)} \|w\|_{L^2(I_i)})\}$$

$$= (bw'', w) + \sum_{i=1}^{M} \bar{b}_i\{\langle w'', w\rangle_i - \int_{I_i} w''w\,dx\} + O(h\|w\|^2_{H^2(I)}).$$

If $p \in P_{r-2}(I)$ and $q \in P_1(I)$, then $r-1 \leq 2r-3$ for $r \geq 2$ and $_1\langle p, q\rangle = (p, q)$. The Peano kernel theorem [5] implies that for $q \in H^2(I)$

$$|_1\langle p, q\rangle - (p, q)| \leq C\|p\|_{L^2(I)} \|q''\|_{L^2(I)}, \quad p \in P_{r-2}(I).$$

Homogeneity shows that

$$|\langle p, q\rangle_i - \int_{I_i} pq\,dx| \leq Ch^2\|p\|_{L^2(I)} \|q''\|_{L^2(I)}, \quad p \in P_{r-2}(I_i).$$

Hence,

$$|\langle w'', w\rangle_i - \int_{I_i} w''w\,dx| \leq C\|w''\|^2_{L^2(I_i)} h^2,$$

and

$$\langle bw'', w\rangle = (bw'', w) + O(h\|w\|^2_{H^2(I)}).$$

Similar references to the Peano kernel theorem and homogeneity show that each of the other terms differs from its corresponding integral by $O(h\|w\|^2_{H^2(I)})$. Therefore, the lemma is demonstrated.

Since $Y - \hat{y} \in \mathcal{M}_1(r, \delta)$, (9.27) and the preceding lemma imply that

(9.28) $$\|L(Y - \hat{y})\|^2_{L^2(I)} \leq \|y\|^2_{W^{2r}(I)} h^{4r-4} + C\|Y - \hat{y}\|^2_{H^2(I)} h.$$

The regularity assumption (9.3) can be applied to (9.28) to give the inequality

$$\|Y - \hat{y}\|^2_{H^2(I)} \leq C\|y\|^2_{W^{2r}(I)} h^{4r-4} + C\|Y - \hat{y}\|^2_{H^2(I)} h.$$

For sufficiently small h,

$$\|Y - \hat{y}\|_{H^2(I)} \leq C\|y\|_{W^{2r}(I)} h^{2r-2},$$

and it follows that

(9.29) $$\|Y - \hat{y}\|_{W^1(I)} \leq C\|y\|_{W^{2r}(I)} h^{2r-2};$$

i.e., the quasi-interpolant $\hat{y}$ differs from the collocation solution Y, along with first derivatives, uniformly by $O(h^{2r-2})$. For $r \geq 3$ this represents a higher order in h than is possible for $y - Y$; however, consider $y - \hat{y}$ at the knots x_i:

(9.30) $$\begin{aligned}(y - \hat{y})(x_i) &= h^{2r-1} D_{r-3}(x_i) = O(\|y\|_{W^{2r-1}(I)} h^{2r-2}),\\ (y - \hat{y})'(x_i) &= h^{2r-1} D'_{r-3}(x_i) = O(\|y\|_{W^{2r-1}(I)} h^{2r-2}).\end{aligned}$$

Therefore, a superconvergence phenomenon occurs at the knots for both the function and its first derivative:

(9.31) $$|(y - Y)^{(k)}(x_i)| \leq C\|y\|_{W^{2r}(I)} h^{2r-2}, \quad i = 0, \ldots, M, \; k = 0, 1.$$

This result is a special case of the deBoor-Swartz development. The proof above can be extended to nonlinear equations, but the object here was to indicate the method of analysis to be used for the parabolic equations of the following sections.

10. Continuous Time Collocation. Let us return to the parabolic problem

$$
\begin{aligned}
(10.1)\qquad & u_t - a(x,t,u)u_{xx} - b(x,t,u,u_x) = 0, \quad x \in I,\ t \in J,\\
& u(0,t) = g_0(t), \quad u(1,t) = g_1(t), \quad t \in J,\\
& u(x,0) = f(x), \quad x \in I,
\end{aligned}
$$

where the coefficients a and b are sufficiently smooth in a neighborhood of the solution u and $0 < m \leq a(x,t,u)$. The collocation procedure to be considered in this section is the same as treated in § 4, except that the initial condition will be modified somewhat in order to obtain the superconvergence at the knots. Let $U: [0,T] \to \mathcal{M}_1(r,\delta)$, $r \geq 4$, satisfy

$$
\begin{aligned}
& \{U_t - a(U)U_{xx} - b(U,U_x)\}(\xi_{ij},t) = 0, \quad i = 1,\dots,M,\ j = 1,\dots,r-1,\\
& \qquad t \in J,\\
(10.2)\qquad & U(0,t) = g_0(t), \quad U(1,t) = g_1(t), \quad t \in J,\\
& U(x,0) = f^*(x), \quad x \in I,
\end{aligned}
$$

where $f^* \in \mathcal{M}_1(r,\delta)$ will be chosen later.

Linearize the differential operator about the solution to define

$$L = \frac{\partial}{\partial t} - A\frac{\partial^2}{\partial x^2} - B\frac{\partial}{\partial x} - G,$$

where

$$
\begin{aligned}
A(x,t) &= a(x,t,u(x,t)),\\
B(x,t) &= b_4(x,t,u(x,t),u_x(x,t)),\\
G(x,t) &= a_3(x,t,u(x,t))u_{xx}(x,t) + b_3(x,t,u(x,t),u_x(x,t)).
\end{aligned}
$$

The analogue of (9.6) reads as follows:

$$L(u - T_{r\delta}u)(\xi_{ij}, t) = \sum_{q=1}^{r-3} \frac{\partial^{r+q+1}}{\partial x^{r+q}\partial t}(x_{i-1/2}, t)C_q(\xi_j)h^{r+q}$$

$$- A(\xi_{ij}, t)\sum_{q=2}^{r-1} \frac{\partial^{r+q}u}{\partial x^{r+q}}(x_{i-1/2}, t)C_q''(\xi_j)h^{r+q-2}$$

(10.3)
$$- B(\xi_{ij}, t)\sum_{q=1}^{r-2} \frac{\partial^{r+q}u}{\partial x^{r+q}}(x_{i-1/2}, t)C_q'(\xi_j)h^{r+q-1}$$

$$- G(\xi_{ij}, t)\sum_{q=1}^{r-3} \frac{\partial^{r+q}u}{\partial x^{r+q}}(x_{i-1/2}, t)C_q(\xi_j)h^{r+q}$$

$$+ E(\xi_{ij}, t)h^{2r-2},$$

where

(10.4) $$E(\xi_{ij}, t) = O(\|u\|_{2r-2,1} + \|u\|_{2r,0}).$$

It will be convenient to introduce the norms

$$\|u\|_{W^k} = \sup_{\alpha+2\beta \leq k} \|u\|_{\alpha,\beta}, \quad k = 0, 1, \ldots .$$

This norm agrees with the W^k-norm introduced in § 9 when applied to functions of x alone. Note that, if a and b are k times boundedly differentiable in a neighborhood of u,

(10.5)
$$\|A\|_{W^k} \leq Q(\|u\|_{W^k} + 1),$$
$$\|B\|_{W^k} \leq Q(\|u\|_{W^{k+1}} + 1),$$
$$\|G\|_{W^k} \leq Q(\|u\|_{W^{k+2}} + 1).$$

Now, expand A, B, and G about a convenient point in I_i, say x_{i-1}, and express $L(u - T_{r\delta}u)$ in the form

$$(10.6)\qquad L(u - T_{r\delta}u)(\xi_{ij}, t) = \sum_{q=0}^{r-3} h^{r+q} R_{iq}^{(0)}(\xi_j, t) + E_i^{(0)}(\xi_j, t) h^{2r-2} ,$$

where (with the first term and the last sum omitted if $q = 0$)

$$(10.7)\qquad \begin{aligned} R_{iq}^{(0)}(\xi_j, t) = {} & \frac{\partial^{r+q+1} u}{\partial x^{r+q} \partial t}(x_{i-1/2}, t) C_q(\xi_j) \\ & - \sum_{\ell=0}^{q} \frac{1}{\ell!} \frac{\partial^\ell A}{\partial x^\ell}(x_{i-1}, t) \frac{\partial^{r+q+2-\ell} u}{\partial x^{r+q+2-\ell}}(x_{i-1/2}, t) \xi_j^\ell C''_{q+2-\ell}(\xi_j) \\ & - \sum_{\ell=0}^{q} \frac{1}{\ell!} \frac{\partial^\ell B}{\partial x^\ell}(x_{i-1}, t) \frac{\partial^{r+q+1-\ell} u}{\partial x^{r+q+1-\ell}}(x_{i-1/2}, t) \xi_j^\ell C'_{q+1-\ell}(\xi_j) \\ & - \sum_{\ell=0}^{q-1} \frac{1}{\ell!} \frac{\partial^\ell G}{\partial x^\ell}(x_{i-1}, t) \frac{\partial^{r+q-\ell} u}{\partial x^{r+q-\ell}}(x_{i-1/2}, t) \xi_j^\ell C_{q-\ell}(\xi_j) . \end{aligned}$$

Then, (2.9), (10.5), and (10.7) yield, for $t \in J$, the constraints

$$(10.8)\qquad \begin{aligned} & {}_1\langle R_{iq}^{(0)}, x^k \rangle = 0, \quad 0 \le k \le r-q-3 , \\ & |R_{iq}^{(0)}(\xi_j, t)| \le K(\|u\|_{W^{r+q+2}}) , \\ & |E_i^{(0)}(\xi_j, t)| \le K(\|u\|_{W^{2r}}) , \end{aligned}$$

where K indicates a generic function of polynomial growth. In fact, the bounds can be seen to be of the form $K(\|u\|_{W^{r+q+1}}) \|u\|_{W^{r+q+2}}$, but for simplicity this refinement will be ignored.

Define recursively corrections $D_0, \ldots, D_{r-4}$ such that $D_k(\cdot, t) \in \mathfrak{M}_1(r, \delta)$ and

$$\tilde{u}^{(\ell+1)} = T_{r,\delta} u + \sum_{k=0}^{\ell} h^{r+k+2} D_k , \quad \tilde{u}^{(0)} = T_{r,\delta} u ,$$

(10.9)

$$L(u-\tilde{u}^{(\ell)})(\xi_{ij}, t) = \sum_{q=\ell}^{r-3} h^{r+q} R_{iq}^{(\ell)}(\xi_j, t) + E_i^{(\ell)}(\xi_j, t) h^{2r-2} ,$$

where

$$h^2 \frac{\partial^2}{\partial x^2} D_k(\xi_{ij}, t) = A(x_{i-1}, t)^{-1} R_{ik}^{(k)}(\xi_j) , \quad i = 1, \ldots, M, \; j = 1, \ldots, r-1,$$

(10. 10)

$$D_k(x_i, t) = \frac{\partial}{\partial x} D_k(x_i, t) = 0, \quad i = 0, \ldots, M.$$

Direct calculation shows that, for $\ell \leq r-4$,

$$\begin{aligned} L(u - \tilde{u}^{(\ell+1)})(\xi_{ij}, t) = & \sum_{q=\ell+1}^{r-3} h^{r+q} R_{iq}^{(\ell)}(\xi_j) + E_i^{(\ell)}(\xi_j) h^{2r-2} - h^{r+\ell+2} \frac{\partial}{\partial t} D_\ell(\xi_{ij}, t) \\ & + h^{r+\ell} \frac{A(\xi_{ij}, t) - A(x_{i-1}, t)}{A(x_{i-1}, t)} R_{i\ell}^{(\ell)}(\xi_j) \\ & + h^{r+\ell+1} B(\xi_{ij}, t) h \frac{\partial}{\partial x} D_\ell(\xi_{ij}, t) + h^{r+\ell+2} G(\xi_{ij}, t) D_\ell(\xi_{ij}, t) \end{aligned} \tag{10. 11}$$

Thus, for $q = \ell+1, \ldots, r-3$,

$$\begin{aligned} R_{iq}^{(\ell+1)}(\xi_j, t) = & R_{iq}^{(\ell)}(\xi_j) + \frac{1}{(q-\ell)!\, A(x_{i-1}, t)} \frac{\partial^{q-\ell} A}{\partial x^{q-\ell}}(x_{i-1}, t) \xi_j^{q-\ell} R_{i\ell}^{(\ell)}(\xi_j) \\ & + \frac{1}{(q-\ell-1)!} \frac{\partial^{q-\ell-1} B}{\partial x^{q-\ell-1}}(x_{i-1}, t) \xi_j^{q-\ell-1} h \frac{\partial}{\partial x} D_\ell(\xi_{ij}, t) \\ & + \frac{1}{(q-\ell-2)!} \frac{\partial^{q-\ell-2} G}{\partial x^{q-\ell-2}}(x_{i-1}, t) \xi_j^{q-\ell-2} D_\ell(\xi_{ij}, t) \\ & - \delta_{q,\ell+2} \frac{\partial}{\partial t} D_\ell(\xi_{ij}, t) , \end{aligned} \tag{10. 12}$$

where $\delta_{q,\ell+2}$ is the Kronecker symbol and the term involving G is omitted for $q = \ell+1$. By induction on ℓ, it can be seen from Lemma 9.2, (10.5), (10.8), and (10.12) that

$$
\begin{aligned}
&{}_1\langle R_{iq}^{(\ell)}, x^k\rangle = 0, \quad 0 \le k \le r-q-3, \quad 0 \le \ell \le r-4, \\
(10.13) \qquad &|R_{iq}^{(\ell)}(\xi_j, t)| \le K(\|u\|_{W^{r+q+2}}), \\
&|E_i^{(\ell)}(\xi_j, t)| \le K(\|u\|_{W^{2r}}).
\end{aligned}
$$

In particular, Lemma 9.2 is used to bound both $h\frac{\partial}{\partial x}D_\ell$ and $\frac{\partial}{\partial t}D_\ell$, since the latter satisfies the equations

$$
\begin{aligned}
&h^2\frac{\partial^2}{\partial x^2}\frac{\partial}{\partial t}D_\ell(\xi_{ij}, t) = \frac{\partial}{\partial t}\frac{1}{A(\xi_{ij}, t)}R_{i\ell}^{(\ell)}(\xi_{ij}, t), \quad i = 1, \dots, M, \quad j = 1, \dots, r-1, \\
(10.14) \qquad & \\
&\frac{\partial}{\partial t}D_\ell(x_i, t) = \frac{\partial}{\partial x}\frac{\partial}{\partial t}D_\ell(x_i, t) = 0, \quad i = 0, \dots, M.
\end{aligned}
$$

The final correction, $D_{r-3}(\cdot, t) \in \mathcal{M}_1(r, \delta)$, can be constructed using Lemma 9.1. Let

$$
\begin{aligned}
&h^2\frac{\partial^2}{\partial x^2}D_{r-3}(\xi_{ij}, t) = A(x_{i-1}, t)^{-1}R_{i,r-3}^{(r-3)}(\xi_j), \quad i = 1, \dots, M, \quad j = 1, \dots, r-1, \\
(10.15) \qquad & \\
&D_{r-3}(0, t) = D_{r-3}(1, t) = 0.
\end{aligned}
$$

Then, the analogue of (10.11) and the a priori bound of Lemma 9.1 imply that, if

$$
(10.16) \qquad \tilde{u} = \tilde{u}^{(r-2)} = T_{r\delta}u + \sum_{k=0}^{r-3} h^{r+k+2}D_k,
$$

$$
(10.17) \qquad \begin{aligned} L(u-\tilde{u})(\xi_{ij},t) &= \rho(\xi_{ij},t), \\ |\rho(\xi_{ij},t)| &\le K(\|u\|_{W^{2r+1}})h^{2r-2}. \end{aligned}
$$

Note that the index on the norm of u appearing in the remainder term has been increased by one; this results from the loss of a factor of h in bounding the non-local function constructed by Lemma 9.1 when the bound is applied to $h^{2r-1}\frac{\partial}{\partial t}D_{r-3}$. It should also be observed that (10.17) holds for $r = 3$ as well; the last correction is also the first one in that case.

Set $\zeta = u-U$ and $\nu = U-\tilde{u}$. It was shown in (6.43), (6.44), and the few lines following them that

$$
(10.18) \qquad (L\nu)(\xi_{ij},t) = \rho(\xi_{ij},t) + (q_1\zeta^2 + q_2\zeta\zeta_x + q_3\zeta\zeta_{xx} + q_4\zeta_x^2)(\xi_{ij},t),
$$

and the q_k's are bounded in a neighborhood of the solution u. The estimate (4.15) yields the bound

$$
(10.19) \qquad \|\zeta\|_{L^\infty(I\times J)} \le C[\|U(0) - T_{r\delta}f\|_{H^1(I)} + h^r\|u\|_{W^{r+4}}].
$$

Now, (4.15) plus homogeneity in h applied to $\mathfrak{M}_1(r,\delta)$ (i.e., "inverse" hypotheses which are valid on $\mathfrak{M}_1(r,\delta)$) show that

$$
(10.20) \qquad \begin{aligned} \|\zeta\|_{W^1} &\le C[h^{-1}\|U(0) - T_{r\delta}f\|_{H^1(I)} + h^{r-1}\|u\|_{W^{r+4}}], \\ |\zeta_{xx}(\xi_{ij},t)| &\le C[\max_{i,j} |U_{xx}(\xi_{ij},0) - (T_{r\delta}f)''(\xi_{ij})| \\ &\quad + h^{-2}\|U(0) - T_{r\delta}f\|_{H^1(I)} + h^{r-2}\|u\|_{W^{r+4}}]. \end{aligned}
$$

The choice of $U(0)$ to be made below will be sufficiently close to $u(0)$ that $\|U(0) - T_{r\delta}f\|_{H^1(I)}$ will be $O(h^{r+1})$; hence, we shall assume that

$$(10.21) \qquad |(q_1\zeta^2 + q_2\zeta\zeta_x + q_3\zeta\zeta_{xx} + q_4\zeta_x^2)(\xi_{ij}, t)| \le K(\|u\|_{W^{2r+1}})h^{2r-2}.$$

The estimate of $\nu(x_i, t)$ can be made under less restrictive hypotheses than that of $\nu_x(x_i, t)$ and will be done first. Let

$$(10.22) \qquad \begin{aligned} L_1 &= \frac{1}{A(x,t)} L = C_1(x,t)\frac{\partial}{\partial t} - \frac{\partial^2}{\partial x^2} - B_1(x,t)\frac{\partial}{\partial x} - G_1(x,t), \\ \rho_1(\xi_{ij}, t) &= \frac{1}{A(\xi_{ij}, t)} (\rho + q_1\zeta^2 + q_2\zeta\zeta_x + q_3\zeta\zeta_{xx} + q_4\zeta_x^2)(\xi_{ij}, t) \\ &= K(\|u\|_{W^{2r+1}})h^{2r-2}. \end{aligned}$$

Take the discrete inner product of $(L_1\nu)(\xi_{ij}, t) = \rho_1(\xi_{ij}, t)$ with the test function ν_t. Then,

$$\begin{aligned} &< C_1\nu_t, \nu_t > - \frac{1}{2}\frac{d}{dt} < \nu_{xx}, \nu > \\ &\qquad = < \rho_1, \nu_t > + < B_1\nu_x, \nu_t > + < G_1\nu, \nu_t >, \end{aligned}$$

and there is a positive constant γ such that

$$\gamma|\nu_t|^2 - \frac{1}{2}\frac{d}{dt} < \nu_{xx}, \nu > \le C(|\rho_1|^2 + |\nu_x|^2 + |\nu|^2).$$

It is a simple consequence of the Gronwall lemma, Lemma 3.3, and the fact that $\nu(0,t) = \nu(1,t) = 0$ that

$$(10.23) \qquad \begin{aligned} &\int_0^T |\nu_t(\tau)|^2 d\tau + \|\nu\|^2_{L^\infty(0,T;H^1(I))} \\ &\qquad \le C\|\nu(0)\|^2_{H^1(I)} + K(\|u\|_{W^{2r+1}})^2 h^{4r-4}. \end{aligned}$$

Thus,

$$(10.24) \qquad |\nu(x_i, t)| \leq C\|\nu(0)\|_{H^1(I)} + K(\|u\|_{W^{2r+1}})h^{2r-2} .$$

At this point, let us make our choice of $U(x, 0)$. Take

$$(10.25) \qquad U(x, 0) = (T_{r\delta}u + \sum_{k=0}^{p} h^{r+k+2}D_k)(x, 0), \quad p = r-4 \text{ or } r-3.$$

It is clear that (10.21) holds and that

$$(10.26) \qquad \|\nu(0)\|_{H^1(I)} \leq \begin{cases} 0 & , \quad p = r-3 \\ K(\|u\|_{W^{2r-1}})h^{2r-2} & , \quad p = r-4. \end{cases}$$

Since

$$(u - \tilde{u})(x_i, t) = h^{2r-1}D_{r-3}(x_i, t) ,$$

then

$$(10.27) \qquad |(u - U)(x_i, t)| \leq K(\|u\|_{W^{2r+1}})h^{2r-2}, \quad i = 0, \ldots, M, \quad t \in J.$$

The following theorem has been proved.

Theorem 10.1. Let the solution u of (10.1) be such that $\|u\|_{W^{2r+1}} < \infty$, where $r \geq 3$. Assume that the coefficients of (10.1) are boundedly differentiable $\max(r-2, 3)$ times in a neighborhood of u. Let $U(x, 0)$ be determined by (10.25) for $p = r-4$ or $p = r-3$ ($p = 0$ for $r = 3$). Let $U(\cdot, t) \in \mathfrak{M}_1(r, \delta)$ be determined by (10.2) for $0 < t \leq T$. Then the estimate (10.27) holds for the error at the knots.

It should be pointed out that $U(x,0)$ can be constructed from the initial function f and the differential equation.

There is more information in (10.23) than we have used. Homogeneity in h again can be studied to yield the bound

$$|\nu_x(x_i,t)| \le K(\|u\|_{W^{2r+1}})h^{2r-5/2} ,$$

assuming (10.25). Thus,

$$(10.28) \qquad |(u-U)_x(x_i,t)| \le K(\|u\|_{W^{2r+1}})h^{2r-5/2} , \quad i=0,\ldots,M,$$

under the hypotheses of Theorem 10.1.

DeBoor-Swartz [6] obtained an $O(h^{2r-2})$ estimate for the error in the first derivative in the two point boundary problem. A modification of the argument above will be given to recover the $h^{1/2}$ lost when homogeneity alone is employed; however, the differential equation will be limited to the linear case with time-independent coefficients:

$$u_t - a(x)u_{xx} - b(x)u_x - c(x)u = F(x,t) , \quad x \in I,\ t \in J,$$

$$(10.29) \qquad u(0,t) = g_0(t) , \quad u(1,t) = g_1(t) , \quad t \in J ,$$

$$u(x,0) = f(x) , \quad x \in I.$$

Let $U:[0,T] \to \mathcal{M}_1(r,\delta)$ be the collocation approximation given by (10.2) and (10.25), $p = r-3$. The operator L coincides with the differential operator of (10.29). The quasi-interpolant u depends in a somewhat simpler fashion on u than in the nonlinear case:

$$(10.30)\qquad \begin{aligned} L(u-\tilde{u})(\xi_{ij},t) &= \rho(\xi_{ij},t) ,\\ |\rho(\xi_{ij},t)| &\le C\|u\|_{W^{2r+1}} h^{2r-2} . \end{aligned}$$

Moreover, the time-derivative of $\tilde{u}$ is the quasi-interpolant of u_t ; hence,

$$(10.31)\qquad |\rho_t(\xi_{ij},t)| \le C\|u\|_{W^{2r+3}} h^{2r-2} .$$

Equation (10.18) simplifies to

$$(10.32)\qquad L\nu(\xi_{ij},t) = \rho(\xi_{ij},t) .$$

Choose $-\nu_{xxt}$ as the test function rather than ν_t. Then,

$$-<\nu_t,\nu_{xxt}> + <a\nu_{xx},\nu_{xxt}> + <b\nu_x,\nu_{xxt}> + <c\nu,\nu_{xxt}>$$
$$= -<\rho,\nu_{xxt}> .$$

Lemma 3.3 implies that

$$\|\nu_t\|^2_{L^2(0,t;H_0^1 I))} + <a\nu_{xx},\nu_{xx}>\Big|_0^t$$
$$= -\int_0^t \{ <\rho,\nu_{xxt}> + <b\nu_x,\nu_{xxt}> + <c\nu,\nu_{xxt}> \}\, d\tau$$
$$= -<\rho + b\nu_x + c\nu,\nu_{xx}>\Big|_0^t + \int_0^t <\rho_t + b\nu_{xt} + c\nu_t,\nu_{xx}>d\tau .$$

Since ν and ν_t vanish on the boundaries $x = 0$ and $x = 1$, it follows that

$$\|\nu_t\|_{L^2(0,T;H^1(I))} + \|\nu_{xx}\|_{L^\infty(0,T;L^2(I))}$$

(10.33) $$\leq C[\|\nu(0)\|_{H^2(I)} + \|u\|_{W^{2r+3}} h^{2r-2}]$$

$$\leq C\|u\|_{W^{2r+3}} h^{2r-2} ,$$

since $\nu(0) = 0$. The vanishing of ν on the boundary combines with (10.33) to give the estimate

$$\|\nu\|_{W^1} \leq C\|u\|_{W^{2r+3}} h^{2r-2} ,$$

by the trivial elliptic regularity associated by the problem $y'' = w$, $x \in I$, $y(0) = y(1) = 0$. Therefore,

(10.34) $$|(u - U)_x(x_i, t)| \leq C\|u\|_{W^{2r+3}} h^{2r-2} .$$

Theorem 10.2. Let $u \in W^{2r+3}$ be the solution of the linear problem (10.29). Assume that the coefficients $a, b,$ and c have at least $\max(r-2, 3)$ bounded derivatives on I. Let U be the solution of (10.2) and (10.25) for $p = r-3$. Then, the first spatial derivative of U approximates that of u at the knots with an error that can be estimated by (10.34).

11. Collocation in Space and Time. Let us return to the study of the collocation methods introduced in § 6. We shall restrict ourselves to the linear problem (with coefficients independent of t) given by

$$
(11.1)\qquad
\begin{aligned}
& Lu = u_t - a(x)u_{xx} - b(x)u_x - c(x)u = 0 \ , \quad x \in I, \ t \in J, \\
& u(0,t) = u(1,t) = 0 \ , \quad t \in J, \\
& u(x,0) = f(x) \ , \quad x \in I \ .
\end{aligned}
$$

Let $U \in \mathfrak{M}_1(r,\delta) \otimes \mathfrak{M}_0(s,\epsilon)$ be determined by the relations

$$
(11.2)\qquad
\begin{aligned}
& \{ U_t - aU_{xx} - bU_x - cU \} (\xi_{ij}, \tau_{k\ell}) = 0, \quad i = 1,\dots,M, \ j = 1,\dots,r-1, \\
& \qquad\qquad\qquad\qquad\qquad\qquad\qquad\qquad k = 1,\dots,N, \ \ell = 1,\dots,s \ , \\
& U(0,t) = U(1,t) = 0 \ , \quad t \in J \ , \\
& U(x,0) = \tilde{u}(x,0) \ , \quad x \in I \ ,
\end{aligned}
$$

where $\tilde{u}(\cdot,t) \in \mathfrak{M}_1(r,\delta)$ is the quasi-interpolant constructed from u and L in §10. We saw in the last section that

$$
(11.3)\qquad
\begin{aligned}
& L(u - \tilde{u})(\xi_{ij}, t) = \rho(\xi_{ij}, t) \ , \quad t \in J, \\
& |\rho(\xi_{ij}, t)| \le C \|u\|_{W^{2r+1}} h^{2r-2} \ .
\end{aligned}
$$

The object of this section is to derive an error estimate at the knots (x_i, t_k) of the form $O(h^{2r-2} + (\Delta t)^{2s})$. While this can be done without making any assumptions regarding relations between r and s and h and Δt, it can be seen after the completion of the argument in the general case (which is merely a messier version of the one to be presented below) that it is natural to assume that

$$(11.4) \qquad r = 2s+1 \ , \quad \Delta t = h^2 \quad (\text{i.e.}, \ h^{2r-2} = (\Delta t)^{2s}) \ ,$$

and we shall make these assumptions throughout this section.

The function $\tilde{u}$ of §10 does not in general lie in $\mathcal{M}_1(r,\delta)\otimes\mathcal{M}_0(s,\epsilon)$, and it is necessary to build a quasi-interpolant in time of $\tilde{u}$ based on L that does lie in space of U in order to apply the analysis that led from (6.25) to (6.27). Recall the interpolation operators $T^*_{s,\epsilon}$, defined by (6.13), and $\mathcal{J} = T_{r,\delta}\otimes T^*_{s,\epsilon}$. Set

$$(11.5) \qquad u^* = T^*_{s,\epsilon}\tilde{u} = \mathcal{J}u + \sum_{\alpha=0}^{r-3} h^{r+\alpha+2} T^*_{s,\epsilon} D_\alpha \ .$$

We wish to represent $L(\tilde{u} - u^*)(\xi_{ij}, \tau_{k\ell})$ in a form useful for modification by corrections in time analogous to the corrections D_α in space.

It follows from (6.11), (6.9), (11.4), and the fact that $C_1^{*\prime}(\tau_\ell) = 0$ that

$$(L(I - T^*_{s\epsilon})T_{r\delta}u)(\xi_{ij}, \tau_{k\ell})$$

$$= \sum_{n=2}^{s} \frac{\partial^{s+n} T_{r\delta}u}{\partial t^{s+n}} (\xi_{ij}, t_{k-1/2}) C_n^{*\prime}(\tau_\ell)(\Delta t)^{s+n-1}$$

$$- \sum_{n=1}^{s-1} \frac{\partial^{s+n}}{\partial t^{s+n}} \left(a\frac{\partial^2}{\partial x^2} + b\frac{\partial}{\partial x} + c\right) T_{r\delta}u(\xi_{ij}, t_{k-1/2}) C_n^*(\tau_\ell)(\Delta t)^{s+n}$$

$$+ O\left(\|T_{r\delta}u\|_{0,2s+1} + \|T_{r\delta}u\|_{1,2s} + \left\|\frac{\partial^2}{\partial x^2} T_{r\delta}u(\xi_{ij},\cdot)\right\|_{0,2s}\right)(\Delta t)^{2s}$$

$$= \sum_{n=1}^{s-1} \frac{\partial^{s+n+1} T_{r\delta} u}{\partial t^{s+n+1}} (\xi_{ij}, t_{k-1/2})(C^{*\prime}_{n+1}(\tau_\ell) - C^*_n(\tau_\ell))(\Delta t)^{s+n}$$

(11.6)

$$+ O(\|u\|_{W^{2r}} + \|T_{r\delta}u\|_{0,2s+1} + \|T_{r\delta}u\|_{1,2s} + \|\frac{\partial^2}{\partial x^2} T_{r\delta}u\|_{0,2s})(\Delta t)^2$$

In shifting $\frac{\partial^{s+n}}{\partial t^{s+n}}(a\frac{\partial^2}{\partial x^2} + b\frac{\partial}{\partial x} + c)T_{r\delta}u$ to $\frac{\partial^{s+n+1}}{\partial t^{s+n+1}} T_{r\delta}u$, it is not necessary in estimating the error induced to use the full exactness of the interpolation operator $T_{r\delta}$ since a factor $(\Delta t)^{s+n}$ is already present. Thus, exactness on polynomials of degree $2(s-n+1)$ leads to (11.6).

The $h^{r+\alpha+2}D_\alpha$ terms can be handled a bit simpler. If we used the full exactness of the interpolation operator $T^*_{s\epsilon}$ (i.e., exact for polynomials of degree s), the leading error terms would be $O((\Delta t)^s h^{r+\alpha+2} + (\Delta t)^{s+1} h^{r+\alpha})$. Both of these are already smaller than $O(h^{2r-2})$; consequently, only remainder terms result from the $h^{r+\alpha+2}D_\alpha$ terms, and not more than $s+1$ differentiations in t are needed on D_α to give the desired estimate. Let $[z]$ denote the greatest integer in z, and set

$$n_1(\alpha) = s+1 - [\frac{(\alpha+3)s}{r-1}] = s - [\frac{\alpha+1}{2}] \text{ , } 0 \le \alpha \le r-3 \text{ ,}$$

(11.7)

$$n_2(\alpha) = s - [\frac{(\alpha+1)s}{r-1}] = n_1(\alpha) \text{ , } 0 \le \alpha \le r-3 \text{ ,}$$

$$n_3(\alpha) = s - [\frac{(\alpha+2)s}{r-1}] = s-1 - [\frac{\alpha}{2}] \text{ , } 0 \le \alpha \le r-3 \text{ .}$$

It is a simple calculation to show that, for $0 \le \alpha \le r-3$ with $\beta(\alpha) = 0$ for $0 \le \alpha \le r-4$ and $\beta(r-3) = 1$,

$$(L(I-T^*_{s\epsilon})h^{r+\alpha+2}D_\alpha)(\xi_{ij},\tau_{k\ell})$$

(11.8) $$= O(\|D_\alpha\|_{0,n_1(\alpha)}h^{\beta(\alpha)} + \|h^2\frac{\partial^2}{\partial x^2}D_\alpha\|_{0,n_2(\alpha)} + \|h\frac{\partial}{\partial x}D_\alpha\|_{0,n_3(\alpha)})h^{2r-2}$$

$$= O(\|u\|_{W^{2r+1}})h^{2r-2} \quad , \quad 0 \le \alpha \le r-3 ,$$

using (10.13)-(10.15), Lemmas 9.1 and 9.2, and the observation that $n_3(\alpha) \le n_2(\alpha) = n_1(\alpha)$. Thus,

$$(L(\tilde{u}-u^*))(\xi_{ij},\tau_{k\ell}) = \sum_{n=1}^{s-1} \frac{\partial^{s+n+1}T_{r\delta}u}{\partial t^{s+n+1}}(\xi_{ij},t_{k-1/2})(C^{*\prime}_{n+1}(\tau_\ell) - C^*_n(\tau_\ell))(\Delta t)^{s+n}$$

(11.9) $$+ \mu^{(0)}(\xi_{ij},\tau_{k\ell})(\Delta t)^{2s} ,$$

where

(11.10) $$|\mu^{(0)}(\xi_{ij},\tau_{k\ell})| \le C\|u\|_{W^{2r+1}} ,$$

since

$$O(\|T_{r\delta}u\|_{0,2s+1} + \|T_{r\delta}u\|_{1,2s} + \|\frac{\partial^2 T_{r\delta}u}{\partial x^2}\|_{0,2s}) = O(\|u\|_{W^{2r+1}}).$$

We shall need an analogue of Lemma 9.2 for the space $\mathcal{M}_0(s,\epsilon)$. Recall the definition (6.10).

Lemma 11.1. Let $0 \le \gamma \le s-1$ and assume that $_1[Q,t^\beta] = 0$ for $\beta = 0,\dots,\gamma$. Then there exists a unique $P \in P_s(I)$ such that

i) $P'(\tau_\ell) = Q(\tau_\ell)$, $\ell = 1,\dots,s$,

ii) $P(0) = P(1) = 0$.

Moreover, $_1[P,t^\beta] = 0$ for $\beta = 0,\dots,\gamma-1$ if $\gamma \ge 1$. Also,

$$\|P\|_{L^\infty(I)} \leq C \max_{\ell=1,\dots,s} |Q(\tau_\ell)|,$$

$$\|P\|_{L^2(I)} \leq C_1[Q].$$

Proof. The polynomial $P' \epsilon P_{s-1}$ is uniquely determined. Set

$$P(t) = \int_0^t P'(\tau)d\tau ;$$

$P(1) = 0$ results from the orthogonality for $\beta = 0$. The orthogonalities for P obviously come from integration by parts and the use of those for Q. The bounds are a consequence of the uniqueness of P.

Let

$$[v, z]_k = \sum_{\ell=1}^{s} v(\tau_{k\ell})z(\tau_{k\ell})w_\ell^* \Delta t ,$$

(11. 11)

$$[v, z] = \sum_{k=1}^{N} [v, z]_k .$$

It is trivial to see that the lemma below follows from the one above.

Lemma 11. 2. Let $0 \leq \gamma \leq s-1$ and let $[Q, t^\beta]_k = 0$, $0 \leq \beta \leq \gamma$, $k = 1, \dots, N$. Then, there exists a unique $P \epsilon \mathcal{M}_0(s, \epsilon)$ such that

i) $\Delta t P'(\tau_{k\ell}) = Q(\tau_{k\ell})$, $k = 1, \dots, N$, $\ell = 1, \dots, s$,

ii) $P(t_k) = 0$, $k = 0, \dots, N$.

Moreover,

$$[P, t^\beta]_k = 0, \quad \beta \leq \gamma - 1 ,$$

$$\|P\|_{L^\infty(J)} \leq C \max_{k,\ell} |Q(\tau_{k\ell})| ,$$

$$\|P\|_{L^2(J)} \leq C[Q, Q]^{1/2} .$$

Note that P is always locally defined; there is no need for an analogue of the non-locally defined function that arose in Lemma 9.1.

The desired quasi-interpolant in $\mathcal{M}_1(r,\delta) \otimes \mathcal{M}_0(s,\epsilon)$ can be constructed from u^* using (11.9) and Lemma 11.2. Note that each of the principal error terms in (11.9) lies in $\mathcal{M}_1(r,\delta)$ for any $\tau_{k\ell}$. Let $D_1^* \in \mathcal{M}_1(r,\delta) \otimes \mathcal{M}_0(s,\epsilon)$ be determined by the equations

$$\Delta t \frac{\partial}{\partial t} D_1^*(x,\tau_{k\ell}) = \frac{\partial^{s+2} T_{r\delta} u}{\partial t^{s+2}}(x, t_{k-1/2})(C_2^{*\prime}(\tau_\ell) - C_1^*(\tau_\ell)),$$

(11.12) $$k = 1,\dots,N, \quad \ell = 1,\dots,s,$$

$$D_1^*(x,t_k) = 0, \quad k = 0,\dots,N.$$

Now, (6.11) and Lemma 11.2 imply that

(11.13) $$[D_1^*(x,\cdot),t^\beta]_k = 0, \quad 0 \le \beta \le s-3,$$

since ${}_1[C_2^{*\prime} - C_1^*, t^\beta] = 0$, $0 \le \beta \le s-2$. Then,

$$(\Delta t)^{s+2}(LD_1^*)(\xi_{ij},\tau_{k\ell}) = (\Delta t)^{s+1} \frac{\partial^{s+2} T_{r\delta} u}{\partial t^{s+2}}(\xi_{ij}, t_{k-1/2})(C_2^{*\prime}(\tau_\ell) - C_1^*(\tau_\ell))$$
$$- (\Delta t)^{s+2}\left(a\frac{\partial^2}{\partial x^2} + b\frac{\partial}{\partial x} + c\right)D_1^*(\xi_{ij},\tau_{k\ell}).$$

We need to express the $(\Delta t)^{s+2}$-term as a functional of $T_{r\delta}u$ and $O((\Delta t)^{2s})$-terms. First, it follows from (11.12) that

$$\Delta t\left(a\frac{\partial^2}{\partial x^2} + b\frac{\partial}{\partial x} + c\right)\frac{\partial}{\partial t} D_1^*(\xi_{ij}, \tau_{k\ell})$$

$$= \frac{\partial^{s+2}}{\partial t^{s+2}}\left(a\frac{\partial^2}{\partial x^2} + b\frac{\partial}{\partial x} + c\right)T_{r\delta}u(\xi_{ij}, t_{k-1/2})(C_2^{*'}(\tau_\ell) - C_1^*(\tau_\ell))$$

$$= \frac{\partial^{s+3}T_{r\delta}u}{\partial t^{s+3}}(\xi_{ij}, t_{k-1/2})(C_2^{*'}(\tau_\ell) - C_1^*(\tau_\ell)) + O(\|u\|_{W^{2r}})(\Delta t)^{s-2} \quad ,$$

again using less than the full exactness of $T_{r\delta}$. Next, denote the solution operator for the equations of Lemma 11. 2 by $\mathcal{K} = \mathcal{K}_{s,\epsilon}$. Then, since $D_1^*(\cdot\,, t_k) = 0$, the application of $\mathcal{K}$ gains one power of Δt and

$$\left(a\frac{\partial^2}{\partial x^2} + b\frac{\partial}{\partial x} + c\right)D_1^{\cdot}(\xi_{ij}, t) = \mathcal{K}\left\{\frac{\partial^{s+3}T_{r\delta}u}{\partial t^{s+3}}(C_2^{*'} - C_1^*)\right\}(\xi_{ij}, t) + O(\|u\|_{W^{2r}})(\Delta t)^{s-1}$$

and

$$\begin{aligned} L(\tilde{u} - u^* - (\Delta t)^{s+2}D_1^*)(\xi_{ij}, \tau_{k\ell}) = {} & [(C_3^{*'}(\tau_\ell) - C_2^*(\tau_\ell))\frac{\partial^{s+3}T_{r\delta}u}{\partial t^{s+3}}(\xi_{ij}, t_{k-1/2}) \\ & + \mathcal{K}\left\{\frac{\partial^{s+3}T_{r\delta}u}{\partial t^{s+3}}(C_2^{*'} - C_1^*)\right\}(\xi_{ij}, \tau_{k\ell})](\Delta t)^{s+2} \\ & + \sum_{n=3}^{s-1}\frac{\partial^{s+n+1}T_{r\delta}u}{\partial t^{s+n+1}}(\xi_{ij}, t_{k-1/2})(C_{n+1}^{*'}(\tau_\ell) - C_n^*(\tau_\ell))(\Delta t)^{s+n} \\ & + O(\|u\|_{W^{2r+1}})(\Delta t)^{2s} \quad . \end{aligned} \tag{11. 14}$$

Lemma 11. 2 shows that both $O((\Delta t)^{s+2})$-terms are orthogonal on each J_k , $k = 1, \ldots, N$, to t^β for $0 \le \beta \le s-3$. Continue by finding

$D_2^* \in \mathfrak{M}_1(r, \delta) \otimes \mathfrak{M}_0(s, \epsilon)$ such that

$$\Delta t \frac{\partial}{\partial \tau} D_2^*(x, \tau_{k\ell}) = (C_3^{*\prime}(\tau_\ell) - C_2^*(\tau_\ell)) \frac{\partial^{s+3} T_{r\delta} u}{\partial t^{s+3}}(x, t_{k-1/2})$$

$$+ \mathcal{K}\{(C_2^{*\prime} - C_1^*) \frac{\partial^{s+3} T_{r\delta} u}{\partial t^{s+3}}\}(x, \tau_{k\ell}),$$

(11.15)

$$k = 1, \ldots, N, \quad \ell = 1, \ldots, s,$$

$$D_2^*(x, t_k) = 0 , \qquad k = 0, \ldots, N.$$

A calculation similar to the one above shows that

$$L(\tilde{u} - u^* - (\Delta t)^{s+2} D_1^* - (\Delta t)^{s+3} D_2^*)(\xi_{ij}, \tau_{k\ell})$$

$$= [(C_4^{*\prime}(\tau_\ell) - C_3^*(\tau_\ell)) \frac{\partial^{s+4} T_{r\delta} u}{\partial t^{s+4}}(\xi_{ij}, t_{k-1/2})$$

(11.16)

$$+ \mathcal{K}\{(C_3^{*\prime} - C_2^*) \frac{\partial^{s+4} T_{r\delta} u}{\partial t^{s+4}}\}(\xi_{ij}, \tau_{k\ell})$$

$$+ \mathcal{K}^2\{(C_2^{*\prime} - C_1^*) \frac{\partial^{s+4} T_{r\delta} u}{\partial t^{s+4}}\}(\xi_{ij}, \tau_{k\ell})](\Delta t)^{s+3}$$

$$+ \sum_{n=4}^{s-1} (C_{n+1}^{*\prime}(\tau_\ell) - C_n^*(\tau_\ell)) \frac{\partial^{s+n+1} T_{r\delta} u}{\partial t^{s+n+1}}(\xi_{ij}, t_{k-1/2})(\Delta t)^{s+n}$$

$$+ O(\|u\|_{W^{2r+1}})(\Delta t)^{2s} .$$

More generally, let $D_m^* \in \mathcal{M}_1(r,\delta) \otimes \mathcal{M}_0(s,\epsilon)$ be given by

(11.17)
$$\Delta t \frac{\partial}{\partial t} D_m(x,\tau_{k\ell}) = \sum_{p=1}^{m-1} \mathcal{K}^p \{ (C^{*\prime}_{m+1-p} - C^*_{m-p}) \frac{\partial^{s+m+1} T_{r\delta} u}{\partial t^{s+m+1}} \}(x,\tau_{k\ell})$$
$$+ (C^{*\prime}_{m+1}(\tau_\ell) - C^*_m(\tau_\ell)) \frac{\partial^{s+m+1} T_{r\delta} u}{\partial t^{s+m+1}}(x,t_{k-1/2}) ,$$
$$k = 1,\ldots,N, \quad \ell = 1,\ldots,s ,$$
$$D_m^*(x,t_k) = 0 , \qquad k = 0,\ldots,N .$$

Let

(11.18)
$$\hat{u} = u^* + \sum_{m=1}^{s-1} (\Delta t)^{s+m+1} D_m^* .$$

Then, it follows that

(11.19)
$$L(\tilde{u} - \hat{u})(\xi_{ij},\tau_{k\ell}) = O(\|u\|_{W^{2r+1}})(\Delta t)^{2s} .$$

Thus, if

$$\nu = U - \hat{u} \in \mathcal{M}_1(r,\delta) \otimes \mathcal{M}_0(s,\epsilon) ,$$

(10.17), (11.2), and (11.19) imply that

(11.20)
$$(L\nu)(\xi_{ij},\tau_{k\ell}) = O(\|u\|_{W^{2r+1}})(\Delta t)^{2s} , \qquad \nu(x,0) = 0 .$$

The argument of §6 can be applied to the function ν. Thus, (11.4), (11.20), (6.25)-(6.27), and Lemma 6.6 show that

(11.21)
$$\|\nu\|_{L^\infty(I\times J)} \le C\|u\|_{W^{2r+1}}(\Delta t)^{2s} .$$

Consequently,

$$(u - U)(x_i, t_k) = (u - \tilde{u})(x_i, t_k) - \nu(x_i, t_k)$$

(11.22) $$= O(\|u\|_{W^{2r+1}})(\Delta t)^{2s}, \quad i = 1, \ldots, M-1, \; k = 0, \ldots, N.$$

Theorem 11. 1. Let u be the solution of the linear problem (11. 1) and let $U \in \mathcal{M}_1(r, \delta) \otimes \mathcal{M}_0(s, \epsilon)$ be the solution of the collocation equation (11. 2). Then, $(u - U)(x_i, t_k) = O(\|u\|_{W^{2r+1}})(\Delta t)^{2s}$ if $r = 2s+1$ and $\Delta t = h^2$.

Appendix.   In section 6 it was necessary to modify the interpolant

$$\mathcal{J}_u \epsilon \mathcal{M} = \mathcal{M}_1(r, \delta) \otimes \mathcal{M}_0(s, \epsilon)$$ by the first spatial correction term of the

full quasi-interpolant.   In order to simplify the argument, r was assumed

to be at least four, so that the first correction required only a locally de-

fined D-term.   For r = 3 no such term exists, and it is necessary to

use the globally constructed correction that comes out of Lemma 9.1.   This

corresponds to the function D_{r-3} constructed in (9.23).   The remainder of

the argument follows as noted in section 6.

CHAPTER III

LOCAL SUPERCONVERGENCE BY LOCAL REFINEMENT

12. Introduction. The knot estimates of Chapter II are suggestive that the solution in a subinterval $I_i = (x_{i-1}, x_i)$ is almost locally obtained in the sense that any modification of the computing procedure on I_i that intuitively should not seriously affect the $O(h^{2r-2})$ error estimate at x_{i-1} and x_i should be permissible without serious effect on the accuracy away from I_i and should produce an accuracy on I_i related to local approximability of the solution of the differential equation. We shall discuss one such modification for both two-point boundary problems and parabolic problems; see [11,12] for related ideas. What we show here is that, if the degree of the polynomial is increased on I_i from r to $p \leq 2r-3$, the uniform error on I_i is improved to become $O(h^{p+1})$. Thus, we receive the full benefit of the knot superconvergence.

13. Local Refinement for the Two-Point Boundary Problem. Consider the two-point boundary problem treated in section 9, and consider the following modification of the collocation procedure (9.2). Let

$$\mathcal{M} = \{v \in C^1(I) \mid v \in P_r(I_k),\ k \neq i;\ v \in P_p(I_i); v(0) = v(1) = 0\}, \tag{13.1}$$

where it will be convenient to choose p such that

(13.2) $$r \le p \le 2r-3.$$

For $k \neq i$, let ξ_{kj}, $j = 1,\dots,r-1$, be exactly as before. For $k = i$, let $\xi_{ij} = x_{i-1} + h\bar{\xi}_j$, $j = 1,\dots,p-1$, where $0 < \bar{\xi}_1 < \bar{\xi}_2 < \dots < \bar{\xi}_{p-1}$ are the Gauss points corresponding to the Legendre polynomial of degree $p-1$. Replace (9.2) by the following: find $Y \in \mathcal{M}$ such that

(13.3) $$LY(\xi_{kj}) = f(\xi_{kj}) \ , \quad j = 1,\dots,r-1 \quad \text{for } k \neq i \quad \text{and}$$
$$j = 1,\dots,p-1 \quad \text{for } k = i \ .$$

The trivial generalization of Lemma 9.3 in which the constant C depends on $\max(r,p) = p$ implies, in conjunction with (9.3), existence and uniqueness of Y for small h. Generalize the interpolation operator $T_{r,\delta}: C^1(I) \to \mathcal{M}_1(r,\delta)$ to the operator $T_h: C^1(I) \cap H_0^1(I) \to \mathcal{M}$ such that

$$(T_h g)(x) = (T_{r,\delta} g)(x) \quad , \quad x \in I \setminus I_i \ ,$$

and $T_h g \in P_p(I_i)$ is determined by

$$(T_h g)^{(\alpha)}(x_\beta) = g^{(\alpha)}(x_\beta) \quad , \quad \alpha = 0,1 \ , \quad \beta = i-1, i \ ,$$
$$(T_h g)(\eta_{ij}^{(p)}) = g(\eta_{ij}^{(p)}) \ , \qquad j = 1,\dots,p-1 \ ,$$

where the superscript p indicates that the roots of $B_p(x) = 0$ should be used in defining $\eta_{ij}^{(p)}$. The argument of (9.4)-(9.8) can be repeated to show that

(13.4) $$L(y - T_h y)(\xi_{ij}) = \sum_{q=0}^{2r-p-3} h^{p+q} R_{iq}^{(0)}(\bar{\xi}_j) + E_i^{(0)}(\bar{\xi}_j) h^{2r-2} \ ,$$

with

$$|R_{iq}^{(0)}(\bar{\xi}_j)| \leq C\|y\|_{W^{p+q+2}(I_i)} ,$$

(13.5) $$|E_i^{(0)}(\bar{\xi}_j)| \leq C\|y\|_{W^{2r}(I_i)} ,$$

$$_1\langle R_{iq}^{(0)}, x^\beta\rangle = \sum_{j=1}^{p-1} R_{iq}^{(0)}(\bar{\xi}_j)\bar{\xi}_j^\beta \bar{w}_j = 0 , \quad 0 \leq \beta \leq p-q-3.$$

Now, carry out the construction of D_α on I_k, $k \neq i$, in exactly the same fashion as in section 9 for $\alpha \leq r-4$. On I_i, it is unnecessary to correct until $\alpha = p-r$; i.e., $D_\alpha(x) = 0$ on I_i for $\alpha = 0, \ldots, p-r-1$. Then, carry forward with the local corrections using Lemma 9.2 on I_i for $\alpha \leq r-4$. For $\alpha = r-3$, construct d_k on I_k and D_{r-3} for $\{d_k\}$ as in the proof of Lemma 9.1; of course, for $p > r$, $d_i'(x_{i-1}) = d_i'(x_i) = 0$. Again, set

(13.6) $$\hat{y} = T_h y + \sum_{\alpha=0}^{r-3} h^{r+\alpha+2} D_\alpha .$$

Then

(13.7) $$L(y-\hat{y})(\xi_{kj}) = \rho(\xi_{kj}) = O(\|y\|_{W^{2r}(I)} h^{2r-2}) ,$$

and it follows as in section 9 that

(13.8) $$\|Y - \hat{y}\|_{H^2(I)} \leq C\|y\|_{W^{2r}(I)} h^{2r-2}$$

for sufficiently small h. Thus,

(13.9) $$\|Y - \hat{y}\|_{W^1} \leq C\|y\|_{W^{2r}(I)} h^{2r-2} .$$

Since

$$\|\hat{y} - T_h y\|_{L^\infty(I_i)} + h\|(\hat{y} - T_h y)'\|_{L^\infty(I_i)} \le C\|y\|_{W^{2r}(I)} h^{\min(2r-2,\,p+2)}$$

and

$$\|y - T_h y\|_{L^\infty(I_i)} + h\|(y-T_h y)'\|_{L^\infty(I_i)} \le C\|y\|_{W^{p+1}(I_i)} h^{p+1} ,$$

then

$$\|y-Y\|_{L^\infty(I_i)} + h\|(y-Y)'\|_{L^\infty(I_i)} \le C\|y\|_{W^{2r}(I)} h^{\min(2r-2,\,p+1)} . \tag{13.10}$$

<u>Theorem 13.1.</u> Let $y \in W^{2r}(I)$ be the solution of (9.1) and let $Y \in \mathfrak{M}$ be the solution of the refined collocation equations (13.3). Then the error $y-Y$ satisfies the local error bound (13.10), $r \le p \le 2r-3$.

Intuitively, the superconvergence has allowed the approximate solution of (9.1) on a single subinterval to reflect optimal order polynomial approximability up through the order of accuracy at the knots. This can produce a significant reduction in computing if the user wishes to know very accurate information only over a small portion of the interval.

Note that the proof works for $p > 2r-3$, but nothing is gained for the additional effort. It is also clear that the proof above can be extended to the case of variable degree r_k in the following way. Let

$$(13.11)\qquad \mathcal{M} = \{v \in C^1(I) \mid v \in P_{r_k}(I_k),\ k = 1, \ldots, M;\ v(0) = v(1) = 0\},$$

and let $\xi_j^{(r)}$, $j = 1, \ldots, r-1$, denote the roots of $B''_r(x) = 0$. Then, set $\xi_{kj} = x_{k-1} + h\xi_j^{(r_k)}$, $j = 1, \ldots, r_k - 1$. Let $Y \in \mathcal{M}$ satisfy

$$(13.12)\qquad (LY)(\xi_{kj}) = f(\xi_{kj}),\quad j = 1, \ldots, r_k - 1,\quad k = 1, \ldots, M.$$

Then, it follows that

$$(13.13)\qquad \|y - Y\|_{L^\infty(I_k)} + h\|(y-Y)'\|_{L^\infty(I_k)} \le C\|y\|_{W^{2r}(I)} h^{\min(2r-2,\, r_k+1)},\quad k = 1, \ldots, M,$$

where

$$(13.14)\qquad r = \min(r_1, \ldots, r_M).$$

An alternative, and perhaps simpler conceptually, method to obtain the available local accuracy involves solving two collocation problems, one global and one local. First, solve the standard collocation problem with a fixed degree for the piecewise-polynomials; i.e., solve (13.12) with $r_k = r$, $k = 1, \ldots, M$. Then, use the boundary values on I_i given by $Y(x_{i-1})$ and $Y(x_i)$ and find a polynomial $Z \in P_p(I_i)$ such that

$$(13.15)\qquad \begin{aligned} &(LZ)(\xi_{ij}) = f(\xi_{ij}),\quad \xi_{ij} = x_{i-1} + h\xi_j^{(p)},\quad j = 1, \ldots, p-1, \\ &Z(x_{i-1}) = Y(x_{i-1}),\ Z(x_i) = Y(x_i). \end{aligned}$$

Such a procedure has been analyzed in detail for the C^0-piecewise polynomial Galerkin process [11]. Results similar to Theorem 13.1 can easily be established.

14. Local Refinement for Parabolic Problems. The local refinement procedure involving the subspace $\mathfrak{M}$ of (13.1) can be extended to parabolic problems. For simplicity, let us consider the linear problem

$$(14.1)\qquad \begin{aligned} Lu = u_t - a(x)u_{xx} - b(x)u_x - c(x)u &= F(x,t), \quad x \in I,\ t \in J, \\ u(0,t) = u(1,t) &= 0, \quad t \in J, \\ u(x,0) &= f(x), \quad x \in I. \end{aligned}$$

Retain the notation of the preceding section and look for a map $U: [0,T] \to \mathfrak{M}$ with $U(x,0)$ to be determined later as a quasi-interpolant of f, such that

$$(14.2)\qquad (LU)(\xi_{jk},t) = F(\xi_{jk},t), \quad \begin{aligned} &j = 1,\dots,M, \\ &k = 1,\dots,r-1, \text{ for } j \neq i, \\ &k = 1,\dots,p-1 \text{ for } j = i. \end{aligned}$$

Then, for $k = 1,\dots,p-1$,

$$\begin{aligned} L(u - T_h u)(\xi_{ik},t) = {} & \sum_{q=1}^{2r-p-3} \frac{\partial^{p+q+1} u}{\partial x^{p+q} \partial t}(x_{i-1/2},t) C_q^{(p)}(\bar{\xi}_k) h^{p+q} \\ & - a(\xi_{ik}) \sum_{q=2}^{2r-p-1} \frac{\partial^{p+q} u}{\partial x^{p+q}}(x_{i-1/2},t) C_q^{(p)\prime\prime}(\bar{\xi}_k) h^{p+q-2} \\ & - b(\xi_{ik}) \sum_{q=1}^{2r-p-2} \frac{\partial^{p+q} u}{\partial x^{p+q}}(x_{i-1/2},t) C_q^{(p)\prime}(\bar{\xi}_k) h^{p+q-1} \\ & - c(\xi_{ik}) \sum_{q=1}^{2r-p-3} \frac{\partial^{p+q} u}{\partial x^{p+q}}(x_{i-1/2},t) C_q^{(p)}(\bar{\xi}_k) h^{p+q} \end{aligned}$$

$$+ O(\|u\|_{W^{2r}} h^{2r-2})$$

(14.3)

$$= \sum_{q=0}^{2r-p-3} R_{iq}^{(0)}(\bar{\xi}_k, t) h^{p+q} + O(\|u\|_{W^{2r}} h^{2r-2}) ,$$

where

$$\sum_{k=1}^{p-1} R_{iq}^{(0)}(\bar{\xi}_k, t) \bar{\xi}_k^{\alpha} \bar{w}_k = 0 \quad , \quad 0 \leq \alpha \leq p-q-3 ,$$

(14.4)

$$|R_{iq}^{(0)}(\bar{\xi}_k, t)| \leq C \|u\|_{W^{p+q+2}} \quad , \quad 0 \leq q \leq 2r-p-3 .$$

The relations (10.6)-(10.8) remain valid for $j \neq i$.

Define the corrections D_α, $\alpha = 0, \ldots, r-4$, in the intervals I_j, $j \neq i$, exactly as in section 10. Similarly, use the proof of Lemma 9.1 to define the pieces $d_{j,r-3}(x,t)$, $j \neq i$, that are combined to form D_{r-3}. Assume that

(14.5) $$r < p \leq 2r-2 .$$

Set

(14.6) $$D_\alpha(x,t) = 0 \quad , \quad x \in I_i, \quad t \in J, \quad 0 < \alpha \leq p-r-1 .$$

Then, use the proof of Lemma 9.1 again to define $d_{i,\alpha}(x,t)$ for $\alpha = p-r, \ldots, r-3$, and note that

(14.7) $$d_{i,\alpha}(x_\beta, t) = \frac{\partial}{\partial x} d_{i,\alpha}(x_\beta, t) = 0 \quad , \beta = i-1, i ,$$

since $p > r$. Form D_{r-3} as in the proof of Lemma 9.1, and set

(14.8) $$\tilde{u} = T_h u + \sum_{\alpha=0}^{r-3} h^{r+\alpha+2} D_\alpha .$$

It is clear that, as in section 10,

$$(14.9)\qquad L(u-\tilde{u})(\xi_{jk},t) = O(\|u\|_{W^{2r+1}}h^{2r-2}) .$$

If $\nu = U - \tilde{u}$, the argument of section 10 implies that

$$(14.10)\qquad \int_0^T |\nu_t|^2 d\tau + \|\nu\|^2_{L^\infty(0,T;H^1(I))} \le C\{\|\nu(0)\|^2_{H^1(I)} + \|u\|^2_{W^{2r+1}}h^{4r-4}\}.$$

Take

$$(14.11)\qquad U(x,0) = \tilde{u}(x,0) ;$$

then

$$(14.12)\qquad \|\nu\|_{L^\infty(I_i\times J)} \le C\|u\|_{W^{2r+1}}h^{2r-2} .$$

On I_i,

$$(u-\tilde{u})(x,t) = (u-T_h u)(x,t) + \sum_{\alpha=p-r}^{r-4} h^{r+\alpha+2} d_{i\alpha}(x,t) + h^{2r-1}D_{r-3}(x,t)$$

$$= O(\|u\|_{W^{2r+1}}h^{\min(p+1,\,2r-2)}) ,$$

since $D_{r-3} = O(h^{-1}\|u\|_{W^{2r+1}})$ and $d_{i\alpha} = O(\|u\|_{W^{p+\alpha+2}})$. Thus,

$$(14.13)\qquad \|u-U\|_{L^\infty(I_i\times J)} \le C\|u\|_{W^{2r+1}}h^{\min(p+1,\,2r-2)} .$$

If $p < 2r-3$, then a slightly modified argument, using fewer corrections in constructing $\tilde{u}$, would allow the norm on u to be taken in the space W^{p+2} in the estimate (14.13). Also since $\partial D_{r-3}/\partial x = O(h^{-1}\|u\|_{W^{2r+1}})$ and

$\partial d_{i\alpha}/\partial x = O(h^{-1}\|u\|_{W^{p+\alpha+2}})$,

$$(14.14)\qquad \left\|\frac{\partial}{\partial x}(u-U)\right\|_{L^{\infty}(I_i\times J)} \le C\|u\|_{W^{2r+1}} h^{\min(p,2r-2)} .$$

Theorem 14.1. Let u be the solution of (14.1), and let U be the solution of (14.2), (14.11). Then the error $u-U$ can be estimated on $I_i\times J$ by (14.13) and (14.14).

It is clear that analogous results can be derived for nonlinear parabolic problems; but, since no new ideas are involved, this argument will be omitted.

The full collocation procedure of section 11 also can be refined; i.e., find $U\in\mathfrak{M}\otimes\mathfrak{M}_0(s,\epsilon)$ such that

$$(14.15)\qquad \begin{aligned} (LU)(\xi_{jk},\tau_{\ell m}) &= F(\xi_{jk},\tau_{\ell m}), \quad j=1,\dots,M,\ k=1,\dots,r-1 \text{ or } p-1,\\ &\qquad \ell=1,\dots,N,\ m=1,\dots,s,\\ U(x,0) &= \tilde{u}(x,0). \end{aligned}$$

The argument of section 11 can be repeated to show that

$$(14.16)\qquad \|u-U\|_{L^{\infty}(I_i\times\{t_0,t_1,\dots,t_N\})} \le C\|u\|_{W^{2r+3}}(h^{\min(p+1,2r-2)}+(\Delta t)^{2s}).$$

If an approximation of u is wanted to very high order on the rectangle $I_i\times J_\ell$, then it is also necessary to refine the subspace $\mathfrak{M}_0(s,\epsilon)$ to use a polynomial of degree $s_1 \leq 2s-1$ on the interval J_ℓ. Then, an uniform estimate of the form $O((\Delta t)^{2s})$ can be obtained on $I_i \times J_\ell$. Note that the calculation is affected on $(I_i \times J) \cup (I \times J_\ell)$.

CHAPTER IV

A SMOOTHED COLLOCATION METHOD AND APPLICATIONS TO EIGENSYSTEM APPROXIMATION

15. Introduction. We shall modify the collocation procedure (9.2) for the two point boundary problem by first projecting the inhomogeneous term f into a piecewise-polynomial space and then obtain more general error estimates than those of §9. In particular, estimates in Sobolev spaces of negative index will be derived; these and some related ones concerning some modified Sobolev spaces will be applied, using the theory of Bramble and Osborn [2, 3], to give eigensystem estimates. The smoothness requirements on the solution of the differential equation will be reduced to the minimal ones needed for approximation when global estimates are sought. Knot superconvergence also is treated, both for the solution of the two point boundary problem and for eigenfunctions. The above results are derived in §16 and §17.

In §18 the smoothed collocation concept is appled to parabolic equations. Again both global estimates and knot superconvergence estimates are found, and the smoothness required of the solution of the differential equation is less than that needed for the standard collocation procedure.

The arguments of this chapter ressemble those employed in analyzing Galerkin methods much more closely than they do the analyses of Chapters I-III. See, for instance, [2, 3, 13, 14, 16].

16. A Smoothed Collocation Method for the Two-Point Boundary Problem.

Consider the usual two-point boundary problem

$$\text{(16.1)}\qquad \begin{aligned} Ly &= y'' + a(x)y' + b(x)y = f(x), \quad x \in I, \\ y(0) &= y(1) = 0, \end{aligned}$$

where for simplicity we shall assume that the coefficients $a(x)$ and $b(x)$ are in $C^{\infty}(I)$. Moreover, assume that there exists a unique solution of (16.1) for each $f \in C(I)$; this implies existence and uniqueness for the adjoint problem $L^{*}\varphi = g$, $x \in I$, $\varphi(0) = \varphi(1) = 0$ and that

$$\text{(16.2)}\qquad \|y\|_{H^{k+2}(I)} \leq C_k \|f\|_{H^k(I)}, \quad k = 0, 1, 2, \ldots .$$

Let $\delta = \{x_0, x_1, \ldots, x_M\}$ as usual and let

$$\text{(16.3)}\qquad h = \max(h_1, \ldots, h_M).$$

Let $Q : L^2(I) \rightarrow \prod_{j=1}^{M} P_{r-2}(I_j)$ denote the standard L^2-projection into polynomials of degree $r-2$ on each subinterval I_j:

$$\text{(16.4)}\qquad (f - Qf, p)_j = \int_{I_j} (f - Qf)\,p\,dx = 0, \ p \in P_{r-2}(I_j), \ Qf \in P_{r-2}(I_j), \ j = 1, \ldots, M.$$

Note that no continuity constraints at the knots are placed on Qf. The smoothed collocation method is then to find $Y \in \mathcal{M} = \mathcal{M}_1(r, \delta) \cap H_0^1(I)$ such that

$$\text{(16.5)}\qquad (LY)(\xi_{ij}) = (Qf)(\xi_{ij}), \quad i = 1, \ldots, M, \ j = 1, \ldots, r-1.$$

The existence and uniqueness of Y again follow from (16.2) and Lemma 9.3 for sufficiently small h, as in section 9. Let $\tilde{y}$ be determined by the relations

$$L\tilde{y} = Qf \ , \quad x \in I,$$

(16.6) $$\tilde{y}(0) = \tilde{y}(1) = 0.$$

Then, (16.2) implies that

$$c\|y - Y\|^2_{H^2(I)} \leq \|L(y - Y)\|^2_{L^2(I)} \leq C[\|y-\tilde{y}\|^2_{H^2(I)} + \|L(\tilde{y}-Y)\|^2_{L^2(I)}]$$

(16.7)

$$= C[\|y-\tilde{y}\|^2_{H^2(I)} + \|Qf - LY\|^2_{L^2(I)}] .$$

First,

(16.8) $$\|y - \tilde{y}\|^2_{H^2(I)} \leq C\sum_{i=1}^{M} \|f - Qf\|^2_{L^2(I_i)} \leq C\sum_{i=1}^{M} h_i^{2t}\|f\|^2_{H^t(I_i)} \ , \quad t \leq r-1,$$

by (16.2) and trivial approximation theory. Since Gaussian quadrature is exact on each I_i for polynomials of degree $2r-3$, the Peano kernel theorem implies that

$$\|Qf - LY\|^2_{L^2(I)} = \|Qf - LY\|^2_{L^2(I)} - <Qf-LY, Qf-LY>$$

(16.9) $$\leq C\sum_{i=1}^{M} h_i^{2r-2}\|\frac{d^{2r-2}}{dx^{2r-2}}(Qf - LY)^2\|_{L^1(I_i)}$$

$$= C\sum_{i=1}^{M} h_i^{2r-2}\|2\{Qf\cdot LY\}^{(2r-2)} + \{(LY)^2\}^{(2r-2)}\|_{L^1(I_i)},$$

since $(Qf)^2 \in P_{2r-4}(I_i)$. Now, it is clear that

$$(16.10)\qquad \begin{aligned} \|\{(LY)^2\}^{(2r-2)}\|_{L^1(I_i)} &\le C\|Y\|^2_{H^r(I_i)}, \\ \|(Qf\cdot LY)^{(2r-2)}\|_{L^1(I_i)} &\le C\|Qf\|_{H^{r-2}(I_i)}\|Y\|_{H^r(I_i)}, \end{aligned}$$

and

$$(16.11)\qquad \|Qf - LY\|^2_{L^2(I)} \le C\sum_{i=1}^{M} h_i^{2r-2}[\|Y\|^2_{H^r(I_i)} + \|Qf\|^2_{H^{r-2}(I_i)}].$$

Next, use Bernstein's theorem [4] and homogeneity on the length of each subinterval possibly to lower the Sobolev space index:

$$(16.12)\qquad \|Qf - LY\|^2_{L^2(I)} \le C\sum_{i=1}^{M} h_i^{2s-2}[\|Y\|^2_{H^s(I_i)} + \|Qf\|^2_{H^{s-2}(I_i)}],\quad 2\le s\le r.$$

For any $w \in P_r(I_i)$,

$$\begin{aligned} h_i^{2s-2}\|Y\|^2_{H^s(I_i)} &\le Ch_i^{2s-2}\{\|Y-w\|^2_{H^s(I_i)} + \|w-y\|^2_{H^s(I_i)} + \|y\|^2_{H^s(I_i)}\} \\ &\le Ch_i^2\|Y-w\|^2_{H^2(I_i)} + Ch_i^{2s-2}\{\|w-y\|^2_{H^s(I_i)} + \|y\|^2_{H^s(I_i)}\} \\ &\le Ch_i^2\{\|Y-y\|^2_{H^2(I_i)} + \|w-y\|^2_{H^2(I_i)}\} \\ &\qquad + Ch_i^{2s-2}\{\|w-y\|^2_{H^s(I_i)} + \|y\|^2_{H^s(I_i)}\}. \end{aligned}$$

Thus, for an appropriate choice of $w \in P_r(I_i)$,

$$(16.13)\qquad h_i^{2s-2}\|Y\|^2_{H^s(I_i)} \le Ch_i^2\|Y - y\|^2_{H^2(I_i)} + Ch_i^{2s-2}\|y\|^2_{H^s(I_i)}.$$

Since the operator Q is bounded on $H^{s-2}(I)$, it follows from (16.9)-(16.13) and (16.2) that

$$\begin{aligned}\|Qf-LY\|^2_{L^2(I)} &\le Ch^2\|Y-y\|^2_{H^2(I)} + C\sum_{i=1}^{M} h_i^{2s-2}(\|f\|^2_{H^{s-2}(I_i)} + \|y\|^2_{H^s(I_i)}) \\ &\le Ch^2\|Y-y\|^2_{H^2(I)} + Ch^{2s-2}\|f\|^2_{H^{s-2}(I)}, \quad 2 \le s \le r.\end{aligned} \tag{16.14}$$

Inequalities (16.7), (16.8), and (16.14) combine to show that

$$c\|Y-y\|^2_{H^2(I)} \le Ch^2\|Y-y\|^2_{H^2(I)} + Ch^{2t}\|f\|^2_{H^t(I)} + Ch^{2s-2}\|f\|^2_{H^{s-2}(I)} \tag{16.15}$$

for $2 \le s \le r$ and $0 \le t \le r-1$. Thus, we have derived an error estimate in $H^2(I)$ for sufficiently small h.

Theorem 16.1. For $h = \max(h_1, \ldots, h_M)$ sufficiently small,

$$\|y - Y\|_{H^2(I)} \le C\|f\|_{H^s(I)} h^s, \quad 0 \le s \le r-1. \tag{16.16}$$

Consider next the error at the knots. Let $G(x, \xi)$ denote the Green's function for (16.1); i.e., for any $\varphi \in H^2(I)$, $\varphi(x) = (G(x,\cdot), L\varphi)$. Then,

$$\begin{aligned}(Y-y)(x_i) &= (G(x_i,\cdot), L(Y-y)) \\ &= (G(x_i,\cdot), Qf - f) + (G(x_i,\cdot), LY - Qf).\end{aligned} \tag{16.17}$$

Smoothness of a and b indicates that $G(x,\cdot)$ is smooth on $[0, x_i]$ and $[x_i, 1]$ and can be approximated as follows:

$$\inf\{\|G(x_i,\cdot) - \chi\|_{L^2(I)} \,|\, \chi \in \prod_{j=1}^{M} P_{r-2}(I_j)\} \le Ch^{r-1}. \tag{16.18}$$

Then, (16.4) implies that

$$(16.19)\qquad \begin{aligned} |(G(x_i,\cdot), Qf-f)| &= |(G(x_i,\cdot) - \chi, Qf - f)|, \quad \chi \in \prod_{j=1}^{M} P_{r-2}(I_j), \\ &\leq Ch^{r+s-1}\|f\|_{H^s(I)}, \quad 0 \leq s \leq r-1. \end{aligned}$$

Also, for $2 \leq s \leq r$,

$$(16.20)\qquad \begin{aligned} |(G(x_i,\cdot), LY - Qf)| &= |(G(x_i,\cdot), LY - Qf) - \langle G(x_i,\cdot), LY - Qf\rangle| \\ &\leq C\sum_{j=1}^{M} \|\{(LY-Qf)\cdot G(x_i,\cdot)\}^{(2r-2)}\|_{L^1(I_j)} h_j^{2r-2} \\ &\leq C\sum_{j=1}^{M} (\|Y\|_{H^r(I_j)} + \|Qf\|_{H^{r-2}(I_j)})\|G\|_{H^{2r-2}(I_j)} h_j^{2r-2} \\ &\leq C\sum_{j=1}^{M} (\|Y\|_{H^s(I_j)} + \|Qf\|_{H^{s-2}(I_j)})\|G\|_{H^{2r-2}(I_j)} h_j^{r+s-2} \\ &\leq C\Big(\sum_{j=1}^{M} (\|Y\|^2_{H^s(I_j)} + \|Qf\|^2_{H^{s-2}(I_j)})h_j^{2s-2}\Big)^{1/2} \cdot \\ &\qquad \cdot \Big(\sum_{j=1}^{M} \|G\|^2_{H^{2r-2}(I_j)} h_j^{2r-2}\Big)^{1/2} \\ &\leq Ch^{r-1}(h^2\|Y-y\|^2_{H^2(I)} + h^{2s-2}\|f\|^2_{H^{s-2}(I)})^{1/2} \end{aligned}$$

using (16.12)-(16.14). Thus, for h sufficiently small,

$$(16.21)\qquad |(G(x_i,\cdot), LY - Qf)| \leq C\|f\|_{H^t(I)} h^{r+t}, \quad 0 \leq t \leq r-2,$$

and

$$(16.22)\qquad |(Y - y)(x_i)| \leq C\|f\|_{H^s(I)} h^{r+s-1}, \quad 0 \leq s \leq r-1.$$

Again we obtain superconvergence at the knots with the rate being $O(h^{2r-2})$ for sufficiently smooth f. Note that the smoothness required is much less than we needed in the argument of section 9; this will be discussed below at greater length.

The estimates (16.16) and (16.22) can be employed to give an error estimate in $L^2(I)$, since

$$\left(\int_{I_i} g(x)^2 dx\right)^{1/2} \leq C\{(|g(x_{i-1})| + |g(x_i)|)h_i^{1/2} + \left(\int_{I_i} g''(x)^2 dx\right)^{1/2} h_i^2\}$$

for any $g \in C^2(I_i)$. Thus,

$$\|Y-y\|_{L^2(I_i)} \leq C\{\|f\|_{H^s(I)} h^{r+s-1} h_i^{1/2} + \|Y-y\|_{H^2(I_i)} h_i^2\}$$

and

$$\|Y-y\|_{L^2(I)} \leq C\{\|f\|^2_{H^s(I)} h^{2r+2s-2} + \|Y-y\|^2_{H^2(I)} h^4\}^{1/2}$$

$$\leq C\|f\|_{H^s(I)} (h^{r+s-1} + h^{s+2}), \quad s \leq r-1 ;$$

i.e.,

$$\text{(16.23)} \qquad \|Y-y\|_{L^2(I)} \leq C\|f\|_{H^{s-1}(I)} h^{s+1}, \quad 1 \leq s \leq r .$$

Standard interpolation theory for operators on Hilbert spaces [18] can be applied to (16.16) and (16.23) to show that

$$\text{(16.24)} \qquad \|Y-y\|_{H^\alpha(I)} \leq C_\alpha \|f\|_{H^s(I)} h^{s+2-\alpha}, \quad 0 \leq \alpha \leq 2,\ 0 \leq s \leq r-1.$$

Note that

$$\text{(16.25)} \qquad \|g\|_{H^\alpha(I)} = \left(\sum_{i=1}^{M} \|g\|^2_{H^\alpha(I_i)}\right)^{1/2} ,$$

with $H^{\alpha}(I_i)$ being given by interpolation by $H^0(I_i) = L^2(I_i)$ and $H^k(I_i)$ for any integer $k > \alpha$, for α non-integral. If we employ this definition for $\alpha > 2$ as well and if we have a quasi-uniform δ (i.e., $\max h_i h_j^{-1} \leq K$ as $h \to 0$), then (16.24) holds for $\alpha \leq s+2$.

Theorem 16.2. For h sufficiently small,

$$\|Y-y\|_{H^{\alpha}(I)} \leq C_{\alpha}\|f\|_{H^s(I)} h^{s+2-\alpha}, \quad 0 \leq \alpha \leq 2, \ 0 \leq s \leq r-1,$$

and

$$|(Y-y)(x_i)| \leq C\|f\|_{H^s(I)} h^{r+s-1}, \quad 0 \leq s \leq r-1, \ i = 1, \ldots, M-1.$$

In order to apply the Bramble-Osborn eigensystem theory we need to obtain certain negative norm estimates. Let

$$\|g\|_{H^{-s}(I)} = \sup_{\psi \in H^s(I)} \frac{(g, \psi)}{\|\psi\|_{H^s(I)}}$$

for $s > 0$ and $g \in L^2(I)$; then complete $L^2(I)$ with respect to this norm to obtain $H^{-s}(I)$ [3]. We shall estimate $(Y-y, \psi)$ for $\psi \in H^s(I)$, $s \leq r-3$. Let $\varphi \in H^{s+2}(I) \cap H_0^1(I)$ be given by the solution of

$$L^*\varphi = \psi, \quad \varphi(0) = \varphi(1) = 0.$$

Then,

$$\begin{aligned} (16.26) \qquad (y - Y, \psi) &= (y - Y, L^*\varphi) = (L(y - Y), \varphi) \\ &= (f - Qf, \varphi) + (Qf - LY, \varphi). \end{aligned}$$

By (16.4) and (16.2) applied to L^*,

$$
\begin{aligned}
|(f - Qf, \varphi)| &= |(f - Qf, \varphi - \chi)| \quad , \quad \chi \in \prod_{i=1}^{M} P_{r-2}(I_i) , \\
\text{(16.27)} \qquad &\leq C \|f\|_{H^{s_1}(I)} \|\varphi\|_{H^{s+2}(I)} h^{s_1+s_2+2} \quad , \quad 0 \leq s \leq r-3, \ 0 \leq s_1 \leq r-1, \\
&\leq C \|f\|_{H^{s_1}(I)} \|\psi\|_{H^{s}(I)} h^{s_1+s+2} .
\end{aligned}
$$

Next, for $\tilde{\varphi} \in \prod_{i=1}^{M} P_{s+1}(I_i)$ chosen appropriately and $0 \leq t \leq r-1$, (16.14) and (16.16) imply that

$$
\begin{aligned}
(Qf - LY, \varphi) &= (Qf - LY, \tilde{\varphi}) + (Qf - LY, \varphi - \tilde{\varphi}) \\
\text{(16.28)} \qquad &= (Qf - LY, \tilde{\varphi}) + O(\|f\|_{H^t(I)} \|\varphi\|_{H^{s+2}(I)} h^{t+s+3}) .
\end{aligned}
$$

By the argument already used several times,

$$
\begin{aligned}
|(Qf - LY, \tilde{\varphi})| &= |(Qf - LY, \tilde{\varphi}) - \langle Qf - LY, \tilde{\varphi} \rangle| \\
&\leq C \sum_{i=1}^{M} \left\| \frac{d^{2r-2}}{dx^{2r-2}} \{(Qf - LY)\tilde{\varphi}\} \right\|_{L^1(I_i)} h_i^{2r-2} \\
\text{(16.29)} \qquad &\leq C \{ \sum_{i=1}^{M} (\|Qf\|^2_{H^{r-2}(I_i)} + \|Y\|^2_{H^r(I_i)}) h_i^{2r-2} \}^{1/2} \\
&\quad \cdot \{ \sum_{i=1}^{M} \|\tilde{\varphi}\|^2_{H^{s+1}(I_i)} h_i^{2r-2} \}^{1/2} .
\end{aligned}
$$

Since $\tilde{\varphi}$ could have been chosen so that both (16.28) and

$$
\|\tilde{\varphi}\|_{H^{s+1}(I)} \leq C \|\varphi\|_{H^{s+1}(I)} \quad , \quad s \leq r-3 ,
$$

are valid, it follows that, for h sufficiently small,

(16.30) $$|(Qf - LY, \varphi)| \le C\|f\|_{H^t(I)} \|\varphi\|_{H^{s+1}(I)} h^{t+r}, \quad 0 \le t \le r-2 .$$

Then, (16.26)-(16.30) imply that

(16.31) $$\begin{aligned} |(y - Y, \psi)| &\le C\|f\|_{H^t(I)} \|\psi\|_{H^s(I)} (h^{s+t+2} + h^{s+t+3} + h^{t+r}) \\ &\le C\|f\|_{H^t(I)} \|\psi\|_{H^s(I)} h^{s+t+2}, \quad 0 \le t \le r-1, 0 \le s \le r-3, \end{aligned}$$

and

(16.32) $$\|y - Y\|_{H^{-s}(I)} \le C\|f\|_{H^t(I)} h^{s+t+2}, \quad 0 \le s \le r-3, \ 0 \le t \le r-1 .$$

In particular,

(16.33) $$\begin{aligned} \|y - Y\|_{H^{-r+3}(I)} &\le C\|f\|_{H^{r-1}(I)} h^{2r-2}, \\ \|y - Y\|_{H^{-r+3}(I)} &\le C\|f\|_{L^2(I)} h^{r-1} . \end{aligned}$$

Theorem 16.3. For h sufficiently small,

(16.34) $$\|y - Y\|_{H^{-s}(I)} \le C\|f\|_{H^t(I)} h^{s+t+2}, \quad 0 \le s \le r-3, \ 0 \le t \le r-1.$$

The H^{-s} estimates above are what is needed for the eigensystem estimates of the next section, but it is also interesting to derive an L^∞ error estimate. Let $\check{y}$ denote the solution of (16.6) as before. The assumption on the differential operator L imply L^∞ elliptic regularity as well as L^2. Thus, elliptic regularity and the Peano kernel theorem imply that

$$c\|y-Y\|_{W^{\infty,2}(I)} \leq \|L(y-Y)\|_{L^{\infty}(I)}$$

$$\leq C\{\|y-\tilde{y}\|_{W^{\infty,2}(I)} + \|L(\tilde{y}-Y)\|_{L^{\infty}(I)}\}$$

(16.35) $$\leq C\{\|f-Qf\|_{L^{\infty}(I)} + \|Qf-LY\|_{L^{\infty}(I)}\}$$

$$\leq C\{\|f\|_{W^{\infty,r-1}(I)} h^{r-1} + \|Qf-LY\|_{L^{\infty}(I)}\}.$$

Since LY interpolates the polynomial $Qf \in P_{r-2}(I_i)$ at $r-1$ points in each each I_i,

(16.36) $$\|Qf-LY\|_{L^{\infty}(I)} \leq C\,h^{r-1} \max_i \left\|\frac{d^{r-1}}{dx^{r-1}} LY\right\|_{L^{\infty}(I_i)} \leq C\|Y\|_{W^{\infty,r}(I)} h^{r-1}.$$

For $\check{y} \in P_r(I_i)$ appropriately chosen,

$$\|Y\|_{W^{\infty,r}(I_i)} \leq \|Y-\check{y}\|_{W^{\infty,r}(I_i)} + \|\check{y}-y\|_{W^{\infty,r}(I_i)} + \|y\|_{W^{\infty,r}(I_i)}$$

(16.37) $$\leq Ch^{2-r}\|Y-\check{y}\|_{W^{\infty,2}(I_i)} + C\|y\|_{W^{\infty,r}(I_i)}$$

$$\leq Ch^{2-r}\|Y-y\|_{W^{\infty,2}(I_i)} + C\|y\|_{W^{\infty,r}(I_i)}.$$

Thus,

(16.38) $$\|Qf-LY\|_{L^{\infty}(I)} \leq C\|y\|_{W^{\infty,r}(I)} h^{r-1} + C\|y-Y\|_{W^{\infty,2}(I)} h$$

and, for h sufficiently small,

(16.39) $$\|y-Y\|_{W^{\infty,2}(I)} \leq C\|f\|_{W^{\infty,r-1}(I)} h^{r-1}.$$

Obviously, the index $r-1$ can be replaced by one between zero and $r-1$. Since it is trivial to see that

$$\|w\|_{L^{\infty}(I_i)} \leq C \max(|w(x_{i-1})|, |w(x_i)|, h^2 \|w\|_{W^{\infty,2}(I_i)}),$$

it follows from the knot estimate (16.22) and (16.39) that

$$(16.40) \qquad \|y-Y\|_{L^{\infty}(I)} \leq C\|f\|_{W^{\infty,s-1}(I)} h^{s+1}, \quad 1 \leq s \leq r.$$

Theorem 16.4. For h sufficiently small, the optimal order L^{∞} estimates (16.39) and (16.40) hold for the error $y-Y$.

Note that the norm applied to f is best possible.

The remainder of this section is devoted to a number of technical lemmas that will be used in the next two sections. The first is related to the the action of the projection Q as an operator on $H^{-s}(I)$.

Lemma 16.5. Let $\beta \in H^t(I)$, $0 \leq t \leq r-1$, and $\varphi \in H^s(I)$, $0 \leq s \leq r-1$. Then

$$|(Q\beta, \varphi)| \leq C \{\|\beta\|_{H^{-s}(I)} + h^{s+t}\|\beta\|_{H^t(I)}\} \|\varphi\|_{H^s(I)}.$$

Proof. For $\chi \in \prod_{i=1}^{M} P_{r-2}(I_i)$ chosen appropriately,

$$\begin{aligned} |(Q\beta, \varphi)| &\leq |(Q\beta-\beta, \varphi-\chi)| + |(\beta, \varphi)| \\ &\leq \|Q\beta-\beta\|_{L^2(I)} \|\varphi-\chi\|_{L^2(I)} + \|\beta\|_{H^{-s}(I)} \|\varphi\|_{H^s(I)} \\ &\leq C\|\beta\|_{H^t(I)} \|\varphi\|_{H^s(I)} h^{s+t} + \|\beta\|_{H^{-s}(I)} \|\varphi\|_{H^s(I)}. \end{aligned}$$

Lemma 16.6. Let $\psi \in \mathfrak{M}$ be the solution of

$$(L\psi)(\xi_{ij}) = (Q\beta)(\xi_{ij}), \quad i = 1, \ldots, M, \quad j = 1, \ldots, r-1,$$

where $\beta \in H^t(I)$, $0 \leq t \leq r-1$. Then, for $0 \leq s \leq r-3$,

$$\|\psi\|_{H^{-s}(I)} \leq C\{\|\beta\|_{H^t(I)} h^{s+t+2} + \|\beta\|_{H^{-s-2}(I)}\}.$$

Proof. Let $\gamma \in H^s(I)$, and let $\varphi \in H^{s+2}(I) \cap H_0^1(I)$ be the solution of $L^* \varphi = \gamma$. Then,

$$\begin{aligned}(\psi, \gamma) &= (\psi, L^*\varphi) = (L\psi, \varphi) \\ &= (Q\beta, \varphi) + (L\psi - Q\beta, \varphi) \\ &= (L\psi - Q\beta, \varphi) + O(\{\|\beta\|_{H^{-s-2}(I)} + \|\beta\|_{H^t(I)} h^{s+t+2}\}\|\varphi\|_{H^{s+2}(I)}),\end{aligned}$$

using Lemma 16.5. If $\tilde{\varphi} \in \prod_{i=1}^{M} P_{s+1}(I_i)$ is chosen properly, then (16.28) and (16.30) show us that

$$\begin{aligned}(L\psi - Q\beta, \varphi) &= (L\psi - Q\beta, \tilde{\varphi}) + (L\psi - Q\beta, \varphi - \tilde{\varphi}) \\ &= O(\|\beta\|_{H^t(I)} \|\varphi\|_{H^{s+1}(I)} h^{t+r}) \\ &\quad + O(\|\beta\|_{H^t(I)} \|\varphi\|_{H^{s+2}(I)} h^{t+s+3}).\end{aligned}$$

Since $\|\varphi\|_{H^{s+2}(I)} \leq C\|\gamma\|_{H^s(I)}$ by the elliptic regularity (16.2), the conclusion of the lemma follows from the estimates above.

<u>Lemma 16.7.</u> For $\psi \in H^1(I)$,

$$\|\psi\|_{L^2(I)} \leq C\{\|Q\psi\|_{L^2(I)} + h\|\psi\|_{H^1(I)}\} .$$

<u>Proof.</u> It is trivial to see that

$$\int_0^1 g(x)^2 dx \leq \left(\int_0^1 g(x)dx\right)^2 + \int_0^1 g'(x)^2 dx , \quad g \in H^1(I) .$$

Let $\tilde{Q}: L^2(I) \to P_{r-2}(I)$ denote the ordinary L^2 projection into polynomials of degree not greater than $r-2$. Clearly,

$$\left(\int_0^1 g(x)dx\right)^2 = \left(\int_0^1 (\tilde{Q}g)(x)dx\right)^2 \leq \int_0^1 (\tilde{Q}g)(x)^2 dx .$$

The lemma then follows from homogeneity in the interval length, applied to each subinterval independently.

17. <u>The Eigensystem Estimates.</u> Theorem 16.3 provides bounds of exactly the nature required by Bramble-Osborn [2, 3]. Let $T: H^s(I) \to H^s(I)$ denote the (compact) map

$$Tf = y, \tag{17.1}$$

where y is the solution of (16.1), and let $T_\delta: L^2(I) \to L^2(I)$ denote the map

$$T_\delta f = Y, \tag{17.2}$$

the solution of (16.5). Let $U_\delta = T - T_\delta$, and set

$$\|U_\delta\|_{\alpha,\beta} = \sup_{\varphi \in H^\beta} \frac{\|U_\delta \varphi\|_{H^\alpha(I)}}{\|\varphi\|_{H^\beta(I)}}, \tag{17.3}$$

where H^γ is to be interpreted by either (16.25) or as the dual of $H^{-\gamma}(I)$. Theorem 16.3 implies the particular bounds

$$\begin{aligned} \|U_\delta\|_{-r+3,\, r-1} &\le ch^{2r-2}, \\ \|U_\delta\|_{-r+3,\, 0} &\le ch^{r-1}, \\ \|U_\delta\|_{0,\, r-1} &\le ch^{r+1}. \end{aligned} \tag{17.4}$$

Consider the eigensystem problems

$$Ty = \lambda y, \quad T_\delta Y = \lambda_\delta Y, \tag{17.5}$$

or, equivalently,

$$\begin{aligned} \lambda L y &= y, \quad x \in I, \\ y(0) &= y(1) = 0, \end{aligned} \tag{17.6}$$

$$(17.7)\qquad \begin{aligned} &\lambda_\delta (LY)(\xi_{ij}) = (QY)(\xi_{ij}) \ , \quad i = 1,\ldots,M, \quad j = 1,\ldots,r-1, \\ &Y(0) = Y(1) = 0 \ . \end{aligned}$$

Let λ be a nonzero eigenvalue of (17.6); its multiplicity is one, since (16.1) is one-dimensional. Let $M(\lambda)$ denote its eigenspace. Then, as h tends to zero, exactly one eigenvalue λ_δ of (17.7) tends to λ; let $M_\delta(\lambda)$ be its eigenspace. It follows from Theorem 3.2 of Bramble-Osborn [3] that

$$(17.8)\qquad \begin{aligned} |\lambda - \lambda_\delta| &\le C\{\|U_h\|_{-r+3,\,r-1} + \|U_h\|_{-r+3,\,0}\|U_h\|_{0,\,r-1} + \|U_h\|^2_{0,\,r-1}\} \\ &\le Ch^{2r-2} \ . \end{aligned}$$

Moreover, there exist eigenfunctions $u(\lambda)$ and $u_\delta(\lambda_\delta)$ of (17.6) and (17.7), respectively, such that

$$(17.9)\qquad \begin{aligned} \|u(\lambda) - u_\delta(\lambda_\delta)\|_{H^{-s}(I)} &\le C\{\|U_h\|_{-s,\,r-1} + \|U_h\|_{-s,0}\|U_h\|_{0,r-1} + \|U_h\|^2_{0,r-1}\} \\ &\le Ch^{r+s+1} \ , \qquad 0 \le s \le r-3 \ , \end{aligned}$$

using [3] and Theorem 16.3.

Let us consider the question of superconvergence at the knots for the eigenfunctions. Let $G(x,\xi)$ denote the Green's function for (16.1). Then, with $u_\delta = u_\delta(\lambda_\delta)$ and $u = u(\lambda)$,

$$(17.10)\qquad \begin{aligned} (u_\delta - u)(x_i) &= (G(x_i,\cdot), L(u_\delta - u)) \\ &= (G(x_i,\cdot), \lambda_\delta Qu_\delta - \lambda u) + (G(x_i,\cdot), Lu_\delta - \lambda_\delta Qu_\delta) \\ &= \lambda_\delta (G(x_i,\cdot), u_\delta - u) + (\lambda_\delta - \lambda)(G(x_i,\cdot), u) \\ &\quad + \lambda_\delta (G(x_i,\cdot), Qu_\delta - u_\delta) + (G(x_i,\cdot), Lu_\delta - \lambda_\delta Qu_\delta). \end{aligned}$$

Since $\lambda_\delta - \lambda = O(h^{2r-2})$, the second term on the right hand side is $O(h^{2r-2})$. Next,

$$(G(x_i,\cdot), Qu_\delta - u_\delta) = (G(x_i,\cdot) - \chi, Qu_\delta - u_\delta), \quad \chi \in \prod_{j=1}^{M} P_{r-2}(I_j),$$

and

$$|(G(x_i,\cdot), Qu_\delta - u_\delta)| \leq Ch^{r-1} \|u_\delta\|_{H^r(I)} h^{r-1} \leq Ch^{2r-2},$$

since (16.25) and (17.9) imply that

$$\begin{aligned}\|u_\delta\|_{H^r(I)} &\leq \|u\|_{H^r(I)} + \|u - u_\delta\|_{H^r(I)} \\ &\leq \|u\|_{H^r(I)} + Ch^{-r}\|u-u_\delta\|_{L^2(I)} \\ &\leq \text{const.}\end{aligned}$$

Also, by the familiar argument of the last section,

$$\begin{aligned}(G(x_i,\cdot), Lu_\delta - \lambda_\delta Qu_\delta) &= (G, Lu_\delta - \lambda_\delta Qu_\delta) - \langle G, Lu_\delta - \lambda_\delta Qu_\delta \rangle \\ &\leq C \sum_{j=1}^{M} \|\{G(Lu_\delta - \lambda_\delta Qu_\delta)\}^{(2r-2)}\|_{L^1(I_j)} h^{2r-2} \\ &\leq C\Big(\sum_{j=1}^{M} \|G\|^2_{H^{2r-2}(I_j)} h^{2r-2}\Big)^{1/2} \Big(\sum_{j=1}^{M} \|u_\delta\|^2_{H^r(I_j)} h^{2r-2}\Big)^{1/2} \\ &\leq Ch^{2r-2}.\end{aligned}$$

Thus,

$$(u_\delta - u)(x_i) = \lambda_\delta(G(x_i,\cdot), u_\delta - u) + O(h^{2r-2}). \tag{17.11}$$

Now, $G(x_i, \cdot) \in H^1(I)$, and (17.9) implies immediately that

$$|(G(x_i, \cdot), u_\delta - u)| \leq C\|u_\delta - u\|_{H^{-1}(I)} \leq Ch^{r+2} ;$$

thus,

(17.12) $$(u_\delta - u)(x_i) = O(h^{r+2}) .$$

We have not used the full power of the negative norm estimates that can be derived using the results of the last section. We shall define a new family of Hilbert spaces and extend the operators T and T_δ to them. Let $\bar{x} \in (0, 1)$ be fixed; we shall assume that we have a sequence of partitions δ such that $\bar{x} = x_{i(\delta)}$ as $h = h(\delta) \to 0$. Let $I' = [0, \bar{x}]$ and $I'' = [\bar{x}, 1]$. For $-\infty < s < \infty$, let

(17.13) $$\tilde{H}^s = H^s(I') \times H^s(I'') \times \mathbb{R}^2 ,$$

and, if $f = (f_1, f_2, \alpha) \in \tilde{H}^s$, set

(17.14) $$\|f\|_s^2 = \|f_1\|^2_{H^s(I')} + \|f_2\|^2_{H^s(I'')} + |\alpha_1|^2 + |\alpha_2|^2 .$$

Note that, for $s \geq 0$, $\tilde{H}^s$ is dense in $\tilde{H}^0$, that, for $s_1 < s_2$, $\tilde{H}^{s_2} \subset \tilde{H}^{s_1}$ and the injection of $\tilde{H}^{s_2}$ into $\tilde{H}^{s_1}$ is compact, and that $\tilde{H}^{-s}$ is the dual of $\tilde{H}^s$, $s > 0$, in the sense that it is the completion of $\tilde{H}^0$ or $C^\infty(I') \times C^\infty(I'') \times \mathbb{R}^2$ with respect to the norm

$$\|\psi\|_{-s} = \sup_{0 \neq \varphi \in \tilde{H}^s} \frac{[\psi, \varphi]}{\|\varphi\|_s} ,$$

where, if $\psi = (\psi_1, \psi_2, \alpha)$ and $\varphi = (\varphi_1, \varphi_2, \beta)$,

$$(17.15)\qquad [\psi,\varphi] = \int_{I'} \psi_1\varphi_1\,dx + \int_{I''} \psi_2\varphi_2\,dx + \alpha_1\beta_1 + \alpha_2\beta_2 .$$

These are the properties required of the family of Hilbert spaces in Theorem 3.2 of Bramble-Osborn [3].

Let us define the operator $T: \tilde{H}^s \to \tilde{H}^{s+2}$. If $f = (f_1, f_2, \alpha) \in \tilde{H}^s$, $s \geq 0$, let

$$(17.16a)\qquad Tf = y = (y_1, y_2, \beta) \in \tilde{H}^{s+2}$$

if and only if

$$(17.16b)\qquad \begin{aligned} Ly_1 &= f_1 , \quad x \in I', \\ Ly_2 &= f_2 , \quad x \in I'', \\ \beta_1 &= y_1(\bar{x}) = y_2(\bar{x}) , \quad \beta_2 = y_1'(\bar{x}) = y_2'(\bar{x}). \end{aligned}$$

Denote the function given on I by y_1 on I' and y_2 on I'' by y^*; then $y^* \in C^1(I)$ for $s \geq 0$. Let us extend the definition of T to negative s in two stages. First, let $s < -2$. If $f \in \tilde{H}^s$, let $f^n = (f_1^n, f_2^n, \alpha^n) \in \tilde{H}^0$ be chosen so that

$$(17.17)\qquad f^n \to f \text{ in } \tilde{H}^s .$$

Let $Tf^n = y^n = (y_1^n, y_2^n, \beta^n) \in \tilde{H}^2$ satisfy (17.16). We want to show that $\{y^n\}$ is Cauchy in $\tilde{H}^{s+2}$. Let $\psi = (\psi_1, \psi_2, \gamma) \in \tilde{H}^{-s-2}$, and $\varphi \in \tilde{H}^{-s}$ satisfy

$$(17.18)\qquad \begin{aligned} L^*\varphi^* &= \psi^* , \quad x \in I, \\ \varphi_1(0) &= \varphi_2(1) = 0, \\ \varphi^* &\in C^1(I) . \end{aligned}$$

Then, if $z = y^n - y^m$,

$$[z, \psi] = (z^*, \psi^*) + z(\bar{x})\gamma_1 + z'(\bar{x})\gamma_2$$
$$= (Lz^*, \varphi^*) + z(\bar{x})\gamma_1 + z'(\bar{x})\gamma_2$$

and

$$|[z, \psi]| \leq (\|Lz\|^2_{H^s(I')} + \|Lz\|^2_{H^s(I'')})^{1/2}(\|\varphi_1\|^2_{H^{-s}(I')} + \|\varphi_2\|^2_{H^{-s}(I'')})^{1/2}$$
$$+ (z(\bar{x})^2 + z'(\bar{x})^2)^{1/2}(\gamma_1^2 + \gamma_2^2)^{1/2} .$$

Now, assume that the regularity (16.2) holds; since we are interested in this section in the eigensystem only, a simple translation of the spectrum by adding a constant to the coefficient $b(x)$ allows us to make this assumption without loss of generality. Thus,

$$|[z, \psi]| \leq C\|f^n - f^m\|_s \|\psi\|_{-s-2} + (z(\bar{x})^2 + z'(\bar{x})^2)^{1/2}(\gamma_1^2 + \gamma_2^2)^{1/2} .$$

A trivial consideration of the Green's function allows us to bound the last term by an expression of the same form as the first term on the right-hand side; hence,

(17.19) $$\|y^n - y^m\|_{s+2} \leq C\|f^n - f^m\|_s \quad , \quad s < -2 ,$$

and $\{y^n\}$ is Cauchy. Thus, the definition of T has been extended to $s < -2$; for $-2 \leq s < 0$, estimate T by interpolation between the maps

$$T: \tilde{H}^{s_0} \to \tilde{H}^{s_0+2} \quad , \quad \text{some } s_0 < -2 ,$$

$$T: \tilde{H}^0 \to \tilde{H}^2 .$$

The choice of the norm on $\tilde{H}^s$ indicates that $y(\bar{x})$ and $y_i'(\bar{x})$ are defined for any s, and, if $y = Tf$,

$$|y_i(\bar{x})| + |y_i'(\bar{x})| \leq C\|f\|_s \quad , \; -\infty < s < \infty .$$

Now, extend T_δ in like manner. It follows from Theorem 16.2 and a simple modification of the proof of Theorem 16.3 that, when the definition (17.3) is interpreted for the $\tilde{H}^s$-spaces,

$$\|T-T_\delta\|_{-s_1, s_2} \leq Ch^{s_1+s_2+2} \quad , \quad 0 \leq s_1 \leq r-3, \quad 0 \leq s_2 \leq r-1. \tag{17.20}$$

Return to the relation (17.11). For $x_i = x_{i(\delta)} = \bar{x}$, it is clear that $G(\bar{x}, \cdot) \in \tilde{H}^s$, any $s > 0$, for smooth $a(x)$ and $b(x)$. Thus

$$(u - u_\delta)(\bar{x}) = O(\|u_\delta - u\|_{-s} + h^{2r-2}) .$$

By Theorem 3.1 of Bramble-Osborn [3], $\|u - u_\delta\|_{-r+3} = O(h^{2r-2})$. and

$$(u - u_\delta)(\bar{x}) = O(h^{2r-2}) . \tag{17.21}$$

The results of this section can be summarized by the following theorem, since a bound for $(u - u_\delta)'(\bar{x})$ can be derived in like fashion.

Theorem 17.1. Let the operator L be defined by (16.1), and let T and T_δ be defined by (17.1) and (17.2), respectively. Let λ and λ_δ be corresponding eigenvalues associated with (17.6) and (17.7), respectively, and let u and u_δ be associated eigenfunctions. Then,

$$|\lambda - \lambda_\delta| \leq Ch^{2r-2} .$$

There exists a selection for u_δ such that, if $\bar{x} = x_{i(\delta)}$,

$$\|u - u_\delta\|_{\tilde{H}^{-s}} \leq Ch^{s+r+1}, \quad 0 \leq s \leq r-3 \ .$$

In particular,

$$|(u-u_\delta)(\bar{x})| + |(u - u_\delta)'(\bar{x})| \leq Ch^{2r-2} \ .$$

18. A Smoothed Collocation Method for Parabolic Equations. Consider the linear parabolic boundary value problem given by

$$(\frac{\partial}{\partial t} - L)u = u_t - a(x)u_{xx} - b(x)u_x - c(x)u = F(x,t), \quad x \in I, \ t \in J,$$

(18.1) $$u(0,t) = u(1,t) = 0 \ , \quad t \in J,$$

$$u(x,0) = f(x) \ , \quad x \in I.$$

Let $\mathcal{M} = \mathcal{M}_1(r,\delta) \cap H_0^1(I)$, $h_i = h$, and let $Q: L^2(I) \to \prod_{i=1}^{M} P_{r-2}(I_i)$ denote the projection introduced in §16. The smoothed collocation method to be studied is the map $U: [0,T] \to \mathcal{M}$ given by the equations

(18.2) $$(\frac{\partial}{\partial t} QU - LU)(\xi_{ij}, t) = (QF)(\xi_{ij}, t) \ , \quad i = 1, \dots, M, \ j = 1, \dots, r-1,$$

with the rule for selecting $U(x,0)$ to be specified later. We shall obtain both global and knot estimates for the error $u - U$. Generalizations to nonlinear equations, introduction of time-discretization, and refinements analogous to those of Chapter III could be carried out without difficulty.

Existence and uniqueness of the solution of (18.2) follows from the fact that $QZ = 0$ and $Z \in \mathcal{M}$ imply that $Z = 0$, which can be shown as follows. First, the orthogonality relations (16.4) necessitate the existence of $r-2$ (counting multiplicities) zeros of Z in the interior of each subinterval I_i. Then, the proof of Lemma 2.3 can be repeated to show that Z vanishes identically.

In order to analyze the error, consider the map $W: [0,T] \to \mathcal{M}$ such that

(18.3) $$(LW)(\xi_{ij}, t) = Q(u_t - F)(\xi_{ij}, t) \ , \quad i = 1, \dots, M, \ j = 1, \dots, r-1, \quad t \in J.$$

If $\eta = W - u$ and $\nu = W - U$, it follows easily that

(18.4) $$(Q\nu_t - L\nu)(\xi_{ij}, t) = Q\eta_t(\xi_{ij}, t).$$

If $L_1 = a(x)^{-1} L$, then

$$<\frac{1}{a} Q\nu_t, Q\nu_t> - <L_1\nu, Q\nu_t> = <Q\eta_t, Q\nu_t>.$$

Now, with $b_1 = ba^{-1}$ and $c_1 = ca^{-1}$,

$$\begin{aligned} <L_1\nu, Q\nu_t> &= (L_1\nu, Q\nu_t) + \{<L_1\nu, Q\nu_t> - (L_1\nu, Q\nu_t)\} \\ &= \frac{1}{2}\frac{d}{dt}\|\nu_x\|^2_{L^2(I)} + (b_1\nu_x + c_1\nu, Q\nu_t) \\ &\quad + O\Big(\sum_{i=1}^{M} \Big\|\frac{\partial^{2r-2}}{\partial x^{2r-2}}\{L_1\nu\cdot Q\nu_t\}\Big\|_{L^1(I_i)} h^{2r-2}\Big), \end{aligned}$$

since $(\nu_{xx}, Q\nu_t) = (\nu_{xx}, \nu_t)$ by (16.4). Homogeneity and Bernstein's theorem imply that

$$\begin{aligned} \sum_{i=1}^{M} \Big\|\frac{\partial^{2r-2}}{\partial x^{2r-2}}\{L_1\nu\cdot Q\nu_t\}\Big\|_{L^1(I_i)} &\le C\Big(\sum_{i=1}^{M}\|\nu\|^2_{H^r(I_i)}\Big)^{1/2}\Big(\sum_{i=1}^{M}\|Q\nu_t\|^2_{H^{r-2}(I_i)}\Big)^{1/2} \\ &\le Ch^{-2r+3}\|\nu\|_{H^1(I)}\|Q\nu_t\|_{L^2(I)}. \end{aligned}$$

Thus, there exists a constant $\rho > 0$ such that

(18.5) $$\rho|Q\nu_t|^2 + \frac{1}{2}\frac{d}{dt}\|\nu\|^2_{H^1_0(I)} \le C\{\|\nu\|^2_{H^1(I)} + \|Q\eta_t\|^2_{L^2(I)}\}.$$

It follows that

(18.6) $$\|\nu\|_{L^\infty(0,T;H^1(I))} \le C\{\|\nu(0)\|_{H^1(I)} + \|Q\eta_t\|_{L^2(0,T;L^2(I))}\}.$$

Since (18.1) and (18.3) show that

$$(L\eta_t)(\xi_{ij}, t) = (Q-I)(u_t - F)_t(\xi_{ij}, t) = (Q-I)Lu_t(\xi_{ij}, t),$$

Theorem 16.2 shows that, as $h \to 0$,

$$\begin{aligned} \|\eta_t(\cdot, t)\|_{L^2(I)} &\le C\|Lu_t(\cdot, t)\|_{H^{r-1}(I)} h^{r+1} \\ &\le C\|u_t(\cdot, t)\|_{H^{r+1}(I)} h^{r+1} . \end{aligned}$$

Thus,

$$\|\nu\|_{L^\infty(0, T; H^1(I))} \le C\{\|\nu(0)\|_{H^1(I)} + \|u_t\|_{L^2(0, T; H^{r+1}(I))} h^{r+1}\} . \tag{18.7}$$

Theorem 16.4 shows that

$$\begin{aligned} \|\eta\|_{L^\infty(I \times J)} &\le C\|u_t - F\|_{L^\infty(0, T; W^{\infty, r-1}(I))} h^{r+1} \\ &\le C\|u\|_{L^\infty(0, T; W^{\infty, r+1}(I))} h^{r+1} . \end{aligned} \tag{18.8}$$

Now, choose $U(x, 0)$ such that

$$\|\nu(0)\|_{H^1(I)} \le C\|u\|_{L^\infty(0, T; W^{\infty, r+1}(I))} h^{r+1} ; \tag{18.9}$$

in particular, the choice $U(x, 0) = W(x, 0)$ given by taking $U(0) \in \mathcal{M}$ to be the solution of

$$(LU)(\xi_{ij}, 0) = (QLf)(\xi_{ij}) , \qquad i = 1, \ldots, M, \ j = 1, \ldots, r-1, \tag{18.10}$$

will suffice for the global bound (18.11) below. Then, (18.7)-(18.9) combine to give the error estimate

$$\|u - U\|_{L^\infty(I \times J)} \le C\{\|u\|_{L^\infty(0, T; W^{\infty, r+1}(I))} + \|u_t\|_{L^2(0, T; H^{r+1}(I))}\} h^{r+1} . \tag{18.11}$$

Theorem 18.1. For h sufficiently small, the error estimate (18.11) holds for the solution of the smoothed collocation equations (18.2), provided that the initial condition U(x, 0) satisfies (18.9).

Notice that the smoothness requirements on u have been lowered somewhat from those of (4.17) in order to obtain optimal order convergence.

Consider next an L^2 estimate for the error. The argument will be some extent parallel an earlier development of one of the authors [16] for Galerkin methods. If $Q\nu$ is used as the test function against (18.4), then

$$<\frac{1}{a} Q\nu_t, Q\nu> - <L_1\nu, Q\nu> = <Q\eta_t, Q\nu> .$$

Note that $<\nu_{xx}, Q\nu> = (\nu_{xx}, Q\nu) = (\nu_{xx}, \nu)$ and $<Q\eta_t, Q\nu> = (Q\eta_t, Q\nu)$ $= (Q\eta_t, \nu)$. Thus,

$$\frac{1}{2}\frac{d}{dt}<\frac{1}{a} Q\nu, Q\nu> + (\nu_x, \nu_x) = (Q\eta_t, \nu) + <b_1\nu_x + c_1\nu, Q\nu>$$

$$\leq \|Q\eta_t\|_{H^{-1}(I)} \|\nu\|_{H^1(I)} + <b_1\nu_x + c_1\nu, Q\nu> . \tag{18.12}$$

Lemma 16.5 and the obvious inequality

$$(|\nu| + |\nu_x|)|Q\nu| \leq \epsilon \|\nu\|^2_{H^1(I)} + C\|Q\nu\|^2_{L^2(I)}$$

show that, for some $\rho > 0$,

$$\rho\frac{d}{dt}<\frac{1}{a} Q\nu, Q\nu> + \|\nu\|^2_{H^1(I)} \leq C\{\|Q\nu\|^2_{L^2(I)} + (\|\eta_t\|_{H^{-1}(I)} + h\|\eta_t\|_{L^2(I)})^2\}.$$

Thus, since $<a^{-1}Q\nu, Q\nu> \geq (\max a(x))^{-1} <Q\nu, Q\nu> = (\max a(x))^{-1}\|Q\nu\|^2_{L^2(I)}$,

$$\| Q\nu\|^2_{L^\infty(0,T;L^2(I))} + \|\nu\|^2_{L^2(0,T;H^1(I))}$$

(18.13)
$$\leq C\{\|Q\nu(0)\|^2_{L^2(I)} + \int_0^T (\|\eta_t\|_{H^{-1}(I)} + h\|\eta_t\|_{L^2(I)})^2 dt\}.$$

Lemma 16.7 can be invoked to produce the inequality

$$\|\nu\|_{L^\infty(0,T;L^2(I))} \leq C\{\|Q\nu(0)\|_{L^2(I)} + \|\eta_t\|_{L^2(0,T;H^{-1}(I))}$$
$$+ h\|\eta_t\|_{L^2(0,T;L^2(I))} + h\|\nu\|_{L^\infty(0,T;H^1(I))}\}.$$

Now, (18.6) and the above give

$$\|\nu\|_{L^\infty(0,T;L^2(I))} \leq C\{\|Q\nu(0)\|_{L^2(I)} + h\|\nu(0)\|_{H^1(I)}$$

(18.14)
$$+ \|\eta_t\|_{L^2(0,T;H^{-1}(I))} + h\|\eta_t\|_{L^2(0,T;L^2(I))}\}.$$

Since $L\eta_t = (Q - I)Lu_t$ at the collocation points, Theorem 16.3 allows us to see that

$$\|\eta_t\|_{L^2(0,T;H^{-1}(I))} + h\|\eta_t\|_{L^2(0,T;L^2(I))}$$

(18.15)
$$\leq C\|Lu_t\|_{L^2(0,T;H^{r-2}(I))} h^{r+1}$$
$$\leq C\|u_t\|_{L^2(0,T;H^r(I))} h^{r+1}.$$

Also, by Theorem 16.3,

(18.16)
$$\|\eta\|_{L^\infty(0,T;L^2(I))} \leq C\|u\|_{L^\infty(0,T;H^{r+1}(I))} h^{r+1}.$$

Now, require that $U(x,0)$ be selected so that

$$\|\nu(0)\|_{L^2(I)} + h\|\nu(0)\|_{H^1(I)} \le C\|f\|_{H^{r+1}(I)} h^{r+1} . \tag{18.17}$$

Almost any rational means of assigning the initial condition will lead to (18.17) being satisfied; in particular, the interpolation operator $T_{r,\delta}$ is sufficient, as well as the procedure given by (18.10). Then, since $u - U = \nu - \eta$, (18.14)-(18.17) imply that

$$\|u - U\|_{L^\infty(0,T;L^2(I))} \le C\{\|u\|_{L^\infty(0,T;H^{r+1}(I))} + \|u_t\|_{L^2(0;T;H^r(I))}\}h^{r+1} \tag{18.18}$$

This estimate agrees in form with the estimates of [16] for the Galerkin method based on the subspace $\mathfrak{M}$ for (18.1). Moreover, it is best possible (at least for odd r) for the case when $F(x,t) = 0$ as is shown in [16].

Theorem 18.2. If the initial condition $U(x,0)$ is selected so that (18.17) holds, then the error $u - U$ can be bounded by the inequality (18.18).

Let us turn to knot estimates. The analysis will again be based on the construction of a quasi-interpolant. We start with $\nu = W - U$ and modify it by smoothed collocation solutions of two point boundary problems. Let $\psi_1 : [0,T] \to \mathfrak{M}$ be determined by the equations

$$(L\psi_1)(\xi_{ij}, t) = (Q\eta_t)(\xi_{ij}, t) , \quad i = 1,\dots,M, \quad j = 1,\dots,n-1, \tag{18.19}$$

and set $\nu_1 = \nu + \psi_1$. Then, (18.4) and (18.19) show that

$$Q\frac{\partial \nu_1}{\partial t} - L\nu_1 = Q\frac{\partial \psi_1}{\partial t} , \quad x = \xi_{ij} , \; t \in J . \tag{18.20}$$

Since $L\psi_{1,t} = Q\eta_{tt}$, Lemma 16.6 implies that

$$\left\|\frac{\partial\psi_1}{\partial t}\right\|_{H^{-s}(I)} \leq C\left\{\left\|\frac{\partial^2\eta}{\partial t^2}\right\|_{H^p(I)} h^{s+p+2} + \left\|\frac{\partial^2\eta}{\partial t^2}\right\|_{H^{-s-2}(I)}\right\},\ 0\leq s\leq r-3, 0\leq p\leq r-1.$$

Since

$$L\frac{\partial^k\eta}{\partial t^k} = (Q-I)L\frac{\partial^k u}{\partial t^k},\quad x = \xi_{ij},$$

Theorems 16.1 and 16.3 show that

$$(18.21)\qquad \left\|\frac{\partial^k\eta}{\partial t^k}\right\|_{H^{-s}(I)} \leq C\left\|L\frac{\partial^k u}{\partial t^k}\right\|_{H^p(I)} h^{s+p+2},\quad -2\leq s\leq r-3,\ 0\leq p\leq r-1.$$

Thus

$$(18.22)\qquad \left\|\frac{\partial\psi_1}{\partial t}\right\|_{H^{-s}(I)} \leq C\left\|L\frac{\partial^2 u}{\partial t^2}\right\|_{H^p(I)} h^{s+p+4},\ 0\leq s\leq r-3,\ 0\leq p\leq r-1.$$

Continue as follows. Let

$$(18.23)\qquad L\psi_{k+1} = Q\frac{\partial\psi_k}{\partial t},\quad x = \xi_{ij},\quad t\in J,$$

and set

$$(18.24)\qquad \nu_k = \nu + \sum_{\ell=1}^{k}\psi_\ell .$$

Then, a simple calculation shows that

$$(18.25)\qquad Q\frac{\partial\nu_k}{\partial t} - L\nu_k = Q\frac{\partial\psi_k}{\partial t},\quad x = \xi_{ij}.$$

Note that, for $x = \xi_{ij}$,

$$L\psi_k = Q\frac{\partial\psi_{k-1}}{\partial t},\quad L\frac{\partial\psi_{k-1}}{\partial t} = Q\frac{\partial^2\psi_{k-2}}{\partial t^2},\ \ldots,\ L\frac{\partial^{k-1}\psi_1}{\partial t^{k-1}} = Q\frac{\partial^k\eta}{\partial t^k}.$$

Appealing to Lemmas 16. 5 and 16. 6 and (18. 22) results in the inequalities

$$\left\|\frac{\partial^k \eta}{\partial t^k}\right\|_{H^{-s}(I)} \leq C \left\| L\frac{\partial^k u}{\partial t^k}\right\|_{H^{r-1}(I)} h^{r+s+1}, \quad 0 \leq s \leq r-3,$$

$$\left\|\frac{\partial^{k-1} \psi_1}{\partial t^{k-1}}\right\|_{H^{-s}(I)} \leq C\left[\left\|\frac{\partial^k \eta}{\partial t^k}\right\|_{L^2(I)} h^{s+2} + \left\|\frac{\partial^k \eta}{\partial t^k}\right\|_{H^{-s-2}(I)}\right]$$

(18. 26) $$\leq C\left\| L\frac{\partial^k u}{\partial t^k}\right\|_{H^{r-1}(I)} h^{r+s+3}, \quad 0 \leq s \leq r-5,$$

$$\left\|\frac{\partial^{k-2} \psi_2}{\partial t^{k-2}}\right\|_{H^{-s}(I)} \leq C\left[\left\|\frac{\partial^{k-1} \psi_1}{\partial t^{k-1}}\right\|_{L^2(I)} h^{s+2} + \left\|\frac{\partial^{k-1} \psi_1}{\partial t^{k-1}}\right\|_{H^{-s-2}(I)}\right]$$

$$\leq C\left\| L\frac{\partial^k u}{\partial t^k}\right\|_{H^{r-1}(I)} h^{r+s+5}, \quad 0 \leq s \leq r-7,$$

. . . .

$$\|\psi_k\|_{H^{-s}(I)} \leq C\left[\left\|\frac{\partial \psi_{k-1}}{\partial t}\right\|_{L^2(I)} h^{s+2} + \left\|\frac{\partial \psi_{k-1}}{\partial t}\right\|_{H^{-s-2}(I)}\right]$$

$$\leq C\left\| L\frac{\partial^k u}{\partial t^k}\right\|_{H^{r-1}(I)} h^{r+s+2k+1}, \quad 0 \leq s \leq r-2k-3 .$$

Let

(18. 27) $$\tilde{\nu} = \nu_k, \quad k = \frac{1}{2}(r-3) \text{ for } r \text{ odd and } k = \frac{1}{2}(r-2) \text{ for } r \text{ even.}$$

For r odd, (18. 26) suffices for the moment. For r even it is necessary to extend (18. 26) so that we can bound ψ_k. In order to do this, we need a modification of Lemma 16. 6.

Lemma 18.3. If $\psi \in \mathfrak{M}$ satisfies $L\psi = Q\beta$ for $x = \xi_{ij}$, then

$$\|\psi\|_{L^2(I)} \le C\{\|\beta\|_{H^{-1}(I)} + h\|\beta\|_{L^2(I)}\} .$$

Proof. Repeat the argument given for Lemma 16.6 with $\gamma \in L^2(I)$ and $\tilde{\varphi} \in \prod_{i=1}^{M} P_1(I_i)$.

We do know that, when $k = \frac{1}{2}(r-2)$ and r is even,

$$\left\|\frac{\partial\psi_{k-1}}{\partial t}\right\|_{H^{-s}(I)} \le C\left\|L\frac{\partial^k u}{\partial t^k}\right\|_{H^{r-1}(I)} h^{2r+s-3} , \quad s = 0, 1 .$$

Thus, the bound

$$\|\psi_k\|_{L^2(I)} \le C\left\|L\frac{\partial^k u}{\partial t^k}\right\|_{H^{r-1}(I)} h^{2r-2} \tag{18.28}$$

holds for r odd or even. We then see from (18.25) that

$$\left\|Q\frac{\partial\tilde{v}}{\partial t}\right\|_{L^2(0,T;L^2(I))} + \|\tilde{v}\|_{L^\infty(0,T;H^1(I))} \le C\left\|L\frac{\partial^{k+1} u}{\partial t^{k+1}}\right\|_{L^2(0,T;H^{r-1}(I))} h^{2r-2}, \tag{18.29}$$

provided that

$$U(0) = W(0) + \sum_{\ell=1}^{k} \psi_\ell(0) , \tag{18.30}$$

which is computable.

Since we know that $(u - W)(x_i, t) = O(h^{2r-2})$, in order to demonstrate that $(u - U)(x_i, t) = O(h^{2r-2})$ it suffices to show that $\psi_\ell(x_i, t) = O(h^{2r-2})$, $i = 0, \dots, M$, $t \in J$. We shall do this by generalizing the duality used in deriving Lemma 16.6. Let $\tilde{H}^s$ denote the space of pairs (ψ, γ), $\gamma \in \mathbb{R}$, such that

$$(18.31)\qquad |||(\psi,\gamma)|||_s^2 = \|\psi\|^2_{H^s(0,\overline{x})} + \|\psi\|^2_{H^s(\overline{x},1)} + \gamma^2 < \infty, \quad s \geq 0,$$

where $\overline{x} = x_i = x_{i(\delta)}$ is fixed as h tends to zero. If $g \in H^1(I)$, define the duality

$$(18.32)\qquad [g,(\psi,\gamma)] = (g,\psi) + g(\overline{x})\gamma, \quad (\psi,\gamma) \in \widetilde{H}^s .$$

Let $\widetilde{H}^{-s}$, $s \geq 0$, be defined by duality with respect to (18.32); we shall need $|||\cdot|||_{-s}$ only for functions in $H^2(I)$. Note that the choice $(\psi,\gamma) = (0,1)$ implies that, when $g \in H^1(I)$,

$$(18.33)\qquad |g(\overline{x})| \leq |||g|||_{-s}, \quad s \geq 0.$$

It can be seen easily that the proof of Lemma 16.5 carries over to $\widetilde{H}^s$ in place of $H^s(I)$, since for $\beta \in H^1(I)$ and $(\varphi,\varphi(\overline{x})) \in \widetilde{H}^s$

$$|(\beta,\varphi)| = |[\beta,\varphi] - \beta(\overline{x})\varphi(\overline{x})| \leq 2|||\beta|||_{-s}|||(\varphi,\varphi(\overline{x}))|||_s .$$

Thus,

$$(18.34)\qquad |(Q\beta,\varphi)| \leq C\{\|\beta\|_{H^t(I)} h^{s+t} + |||\beta|||_{-s}\}|||(\varphi,\varphi(\overline{x}))|||_s, \quad 0 \leq s \leq r-1.$$

Lemma 16.6 can be extended as follows.

<u>Lemma 18.4.</u> Let $\psi \in \mathcal{M}$ be the solution of

$$(L\psi)(\xi_{ij}) = (Q\beta)(\xi_{ij}), \quad i = 1,\dots,M,\ j = 1,\dots,r-1,$$

when $\beta \in H^t(I)$, $0 \leq t \leq r-1$. Then

$$(18.35)\qquad |||\psi|||_{-s} \leq C\{\|\beta\|_{H^t(I)} h^{s+t+2} + |||\beta|||_{-s-2}\}, \quad 0 \leq s \leq r-3.$$

Proof. The demonstration is patterned after that of Lemma 16.6. Let $(\mu, \gamma) \in \tilde{H}^s$ and define φ by the relations

$$L^* \varphi = (a\varphi)'' - (b\varphi)' + c\varphi = \mu \ , \quad x \in I \backslash \{\bar{x}\},$$

$$\varphi(0) = \varphi(1) = 0,$$

$$\left(a \frac{d\varphi}{dx}\right)\Big|_{\bar{x}-0}^{\bar{x}+0} = \gamma \ .$$

Then, it is easy to see that

$$[\psi, (\mu, \gamma)] = (L\psi, \varphi) = (Q\beta, \varphi) + (L\psi - Q\beta, \varphi) \ .$$

Now, if we assume that $(Lu, u) \geq \rho \|u\|^2_{H^1(I)}$ for some $\rho > 0$ (and this, at most, amounts to a change in the time scale) and that the coefficients are smooth enough, then it follows that there exists a constant C such that

$$|||(\varphi, \varphi(\bar{x}))|||_{s+2} \leq C |||(\mu, \gamma)|||_s \ .$$

The remainder of the proof remains essentially unaltered, and (18.35) results.

Next, let us show that

$$(18.36) \qquad |||\frac{\partial^\ell \eta}{\partial t^\ell}|||_{-s} \leq C \| L \frac{\partial^\ell u}{\partial t^\ell} \|_{H^{r-1}(I)} h^{r+s+1} , \quad 0 \leq s \leq r-3.$$

With (μ, γ) and φ as above,

$$\left[\frac{\partial^\ell \eta}{\partial t^\ell}, (\mu, \gamma)\right] = \left(L \frac{\partial^\ell \eta}{\partial t^\ell}, \varphi\right) = \left(L \frac{\partial^\ell w}{\partial t^\ell} - QL \frac{\partial^\ell u}{\partial t^\ell}, \varphi\right) - \left((I-Q)L \frac{\partial^\ell u}{\partial t^\ell}, \varphi\right)$$

$$= O\left(\| L \frac{\partial^\ell u}{\partial t^\ell} \|_{H^{r-1}(I)} |||(\mu, \gamma)|||_s h^{r+s+1}\right) ,$$

using (16. 14), (16. 16), (16. 28), and (16. 30). Hence, (18. 36) holds. An argument analogous to the one leading to (18. 26) shows that

$$(18.37)\qquad |||\frac{\partial^n \psi_\ell}{\partial t^n}|||_{-s} \leq C \| L \frac{\partial^{\ell+n} u}{\partial t^{\ell+n}} \|_{H^{r-1}(I)} h^{r+s+2\ell+1}$$

for $0 \leq s \leq r-2\ell-3$ and that, for r odd or even,

$$(18.38)\qquad |||\frac{\partial^n \psi_k}{\partial t^n}|||_0 \leq C \| L \frac{\partial^{k+n} u}{\partial t^{k+n}} \|_{H^{r-1}(I)} h^{2r-2} .$$

In particular,

$$(18.39)\qquad \|\psi_\ell(\bar{x},\cdot)\|_{L^\infty(0,T)} \leq C \| L \frac{\partial^k u}{\partial t^k} \|_{L^\infty(0,T;\, H^{r-1}(I))} h^{2r-2}, \quad \ell = 0,\dots,k.$$

Since $(u-U)(x_i,t) = (u-W)(x_i,t) + \tilde{\nu}(x_i,t) - \psi_1(x_i,t) - \dots - \psi_k(x_i,t)$, Theorem 16. 2, (18. 29), and (18. 39) combine to complete the proof of the following statement.

<u>Theorem 18. 5.</u> The error in the approximate solution U of (18. 2) at a knot $\bar{x} = x_{i(\delta)}$ satisfies the inequality

$$(18.40)\qquad |(u-U)(\bar{x},t)| \leq C\{\|L\frac{\partial^k u}{\partial t^k}\|_{L^\infty(0,T;H^{r-1}(I))} + \|L\frac{\partial^{k+1} u}{\partial t^{k+1}}\|_{L^2(0,T;H^{r-1}(I))}\} h^{2r-2},$$

where $k = \frac{1}{2}(r-3)$ for r odd and $k = \frac{1}{2}(r-2)$ for r even and $U(0)$ is determined by (18. 30).

For $F = 0$, the bound can stated in the form

$$\|(u-U)(\bar{x},\cdot)\|_{L^{\infty}(0,T)} \leq C\|u\|_{L^2(0,T;H^{r+2k+3}(I))} h^{2r-2}$$

(18.41)

$$= \begin{cases} C\|u\|_{L^2(0,T;H^{2r}(I))} h^{2r-2}, & r \text{ odd}, \\ C\|u\|_{L^2(0,T;H^{2r+1}(I))} h^{2r-2}, & r \text{ even}. \end{cases}$$

References

1. D. Archer, Thesis, Rice University, 1973.

2. J.H. Bramble and J.E. Osborn, Rate of convergence estimates for nonselfadjoint eigenvalue approximations, Math. Comp. 27(1973), 525-549.

3. _____, _____, Approximation of Steklov eigenvalues of nonselfadjoint second order elliptic operators, The Mathematical Foundations of the Finite Element Method with Applications to Partial Differential Equations, A.K. Aziz (ed.), Academic Press, New York, 1972.

4. E.W. Cheney, Introduction to Approximation Theory, McGraw-Hill, New York, 1966.

5. P.J. Davis, Interpolation and Approximation, Blaisdell Publishing Co., New York, 1963.

6. C. DeBoor and B. Swartz, Collocation at Gaussian points, SIAM J. Numer. Anal. 10(1973), 582-606.

7. J. Douglas, Jr., A superconvergence result for the approximation solution of the heat equation by a collocation method, The Mathematical Foundations of the Finite Element Method with Applications to Partial Differential Equations, A.K. Aziz (ed.), Academic Press, New York, 1972.

8. J. Douglas, Jr., and T. Dupont, A finite element collocation method for the heat equation, Symposia Mathematica 10, Monograf, Bologna, 1972.

9. _____, _____, A finite element collocation method for quasilinear parabolic equations, Math. Comp. 27, (1973), 17-28.

10. _____, _____, Some superconvergence results for Galerkin methods for the approximate solution of two point boundary problems, to appear in the Proceedings of a conference on numerical analysis held by the Royal Irish Academy, Dublin, 1972.

11. _____, _____, Galerkin approximations for the two point boundary problem using continuous, piecewise polynomial spaces, to appear in Numer. Math.

12. _____, _____, Superconvergence for Galerkin methods for the two point boundary problem via local projections, to appear in Numer. Math.

13. _____, _____, and M.F. Wheeler, Some superconvergence results for an H^1-Galerkin procedure for the heat equation, MRC Report No. 1382 and to appear in the Proceedings of an International Symposium on Computing Methods in Applied Sciences and Engineering, IRIA, Rocquencourt, 1973.

14. J. Douglas, Jr., T. Dupont, and M.F. Wheeler, A quasi-projection approximation method applied to Galerkin procedures for parabolic and hyperbolic equations, to appear.

15. _____, _____, _____, A Galerkin procedure for approximating the flux on the boundary for elliptic and parabolic boundary value problems MRC Report No. 1381 and to appear.

16. T. Dupont, Some L^2 error estimates for parabolic Galerkin methods, The Mathematical Foundations of the Finite Element Method with Applications to Partial Differential Equations, A.K. Aziz (ed.), Academic Press, New York, 1972.

17. B.L. Hulme, One-step piecewise polynomial Galerkin methods for initial value problems, Math. Comp. 26 (1972), 415-426.

18. J.-L. Lions and E. Magenes, Problèmes aux limites non homogènes et applications, vol. 1, Dunod, Paris, 1968.

19. T.R. Lucas and G.W. Reddien, Some collocation methods for nonlinear boundary value problems, SIAM J. Numer. Anal. 9(1972), 341-356.

20. V. Thomée, Spline approximation and difference schemes for the heat equation, The Mathematical Foundations of the Finite Element Method with Applications to Partial Differential Equations, A.K. Aziz (ed.), Academic Press, New York, 1972.

21. V. Thomée, and B. Wendroff, Convergence estimates for Galerkin methods for variable coefficient initial-value problems, to appear.

22. E. T. Whittaker and G. N. Watson, A Course in Modern Analysis, Cambridge University Press, New York, 1948.

Vol. 215: P. Antonelli, D. Burghelea and P. J. Kahn, The Concordance-Homotopy Groups of Geometric Automorphism Groups. X, 140 pages. 1971. DM 16,-

Vol. 216: H. Maaß, Siegel's Modular Forms and Dirichlet Series. VII, 328 pages. 1971. DM 20,-

Vol. 217: T. J. Jech, Lectures in Set Theory with Particular Emphasis on the Method of Forcing. V, 137 pages. 1971. DM 16,-

Vol. 218: C. P. Schnorr, Zufälligkeit und Wahrscheinlichkeit. IV, 212 Seiten. 1971. DM 20,-

Vol. 219: N. L. Alling and N. Greenleaf, Foundations of the Theory of Klein Surfaces. IX, 117 pages. 1971. DM 16,-

Vol. 220: W. A. Coppel, Disconjugacy. V, 148 pages. 1971. DM 16,-

Vol. 221: P. Gabriel und F. Ulmer, Lokal präsentierbare Kategorien. V, 200 Seiten. 1971. DM 18,-

Vol. 222: C. Meghea, Compactification des Espaces Harmoniques. III, 108 pages. 1971. DM 16,-

Vol. 223: U. Felgner, Models of ZF-Set Theory. VI, 173 pages. 1971. DM 16,-

Vol. 224: Revètements Etales et Groupe Fondamental. (SGA 1). Dirigé par A. Grothendieck XXII, 447 pages. 1971. DM 30,-

Vol. 225: Théorie des Intersections et Théorème de Riemann-Roch. (SGA 6). Dirigé par P. Berthelot, A. Grothendieck et L. Illusie. XII, 700 pages. 1971. DM 40,-

Vol. 226: Seminar on Potential Theory, II. Edited by H. Bauer. IV, 170 pages. 1971. DM 18,-

Vol. 227: H. L. Montgomery, Topics in Multiplicative Number Theory. IX, 178 pages. 1971. DM 18,-

Vol. 228: Conference on Applications of Numerical Analysis. Edited by J. Ll. Morris. X, 358 pages. 1971. DM 26,-

Vol. 229: J. Väisälä, Lectures on n-Dimensional Quasiconformal Mappings. XIV, 144 pages. 1971. DM 16,-

Vol. 230: L. Waelbroeck, Topological Vector Spaces and Algebras. VII, 158 pages. 1971. DM 16,-

Vol. 231: H. Reiter, L^1-Algebras and Segal Algebras. XI, 113 pages. 1971. DM 16,-

Vol. 232: T. H. Ganelius, Tauberian Remainder Theorems. VI, 75 pages. 1971. DM 16,-

Vol. 233: C. P. Tsokos and W. J. Padgett. Random Integral Equations with Applications to stochastic Systems. VII, 174 pages. 1971. DM 18,-

Vol. 234: A. Andreotti and W. Stoll. Analytic and Algebraic Dependence of Meromorphic Functions. III, 390 pages. 1971. DM 26,-

Vol. 235: Global Differentiable Dynamics. Edited by O. Hájek, A. J. Lohwater, and R. McCann. X, 140 pages. 1971. DM 16,-

Vol. 236: M. Barr, P. A. Grillet, and D. H. van Osdol. Exact Categories and Categories of Sheaves. VII, 239 pages. 1971. DM 20,-

Vol. 237: B. Stenström, Rings and Modules of Quotients. VII, 136 pages. 1971. DM 16,-

Vol. 238: Der kanonische Modul eines Cohen-Macaulay-Rings. Herausgegeben von Jürgen Herzog und Ernst Kunz. VI, 103 Seiten. 1971. DM 16,-

Vol. 239: L. Illusie, Complexe Cotangent et Déformations I. XV, 355 pages. 1971. DM 26,-

Vol. 240: A. Kerber, Representations of Permutation Groups I. VII, 192 pages. 1971. DM 18,-

Vol. 241: S. Kaneyuki, Homogeneous Bounded Domains and Siegel Domains. V, 89 pages. 1971. DM 16,-

Vol. 242: R. R. Coifman et G. Weiss, Analyse Harmonique Non-Commutative sur Certains Espaces. V, 160 pages. 1971. DM 16,-

Vol. 243: Japan-United States Seminar on Ordinary Differential and Functional Equations. Edited by M. Urabe. VIII, 332 pages. 1971. DM 26,-

Vol. 244: Séminaire Bourbaki - vol. 1970/71. Exposés 382-399. IV, 356 pages. 1971. DM 26,-

Vol. 245: D. E. Cohen, Groups of Cohomological Dimension One. V, 99 pages. 1972. DM 16,-

Vol. 246: Lectures on Rings and Modules. Tulane University Ring and Operator Theory Year, 1970-1971. Volume I. X, 661 pages. 1972. DM 40,-

Vol. 247: Lectures on Operator Algebras. Tulane University Ring and Operator Theory Year, 1970-1971. Volume II. XI, 786 pages. 1972. DM 40,-

Vol. 248: Lectures on the Applications of Sheaves to Ring Theory. Tulane University Ring and Operator Theory Year, 1970-1971. Volume III. VIII, 315 pages. 1971. DM 26,-

Vol. 249: Symposium on Algebraic Topology. Edited by P. J. Hilton. VII, 111 pages. 1971. DM 16,-

Vol. 250: B. Jónsson, Topics in Universal Algebra. VI, 220 pages. 1972. DM 20,-

Vol. 251: The Theory of Arithmetic Functions. Edited by A. A. Gioia and D. L. Goldsmith VI, 287 pages. 1972. DM 24,-

Vol. 252: D. A. Stone, Stratified Polyhedra. IX, 193 pages. 1972. DM 18,-

Vol. 253: V. Komkov, Optimal Control Theory for the Damping of Vibrations of Simple Elastic Systems. V, 240 pages. 1972. DM 20,-

Vol. 254: C. U. Jensen, Les Foncteurs Dérivés de $\varprojlim$ et leurs Applications en Théorie des Modules. V, 103 pages. 1972. DM 16,-

Vol. 255: Conference in Mathematical Logic - London '70. Edited by W. Hodges. VIII, 351 pages. 1972. DM 26,-

Vol. 256: C. A. Berenstein and M. A. Dostal, Analytically Uniform Spaces and their Applications to Convolution Equations. VII, 130 pages. 1972. DM 16,-

Vol. 257: R. B. Holmes, A Course on Optimization and Best Approximation. VIII, 233 pages. 1972. DM 20,-

Vol. 258: Séminaire de Probabilités VI. Edited by P. A. Meyer. VI, 253 pages. 1972. DM 22,-

Vol. 259: N. Moulis, Structures de Fredholm sur les Variétés Hilbertiennes. V, 123 pages. 1972. DM 16,-

Vol. 260: R. Godement and H. Jacquet, Zeta Functions of Simple Algebras. IX, 188 pages. 1972. DM 18,-

Vol. 261: A. Guichardet, Symmetric Hilbert Spaces and Related Topics. V, 197 pages. 1972. DM 18,-

Vol. 262: H. G. Zimmer, Computational Problems, Methods, and Results in Algebraic Number Theory. V, 103 pages. 1972. DM 16,-

Vol. 263: T. Parthasarathy, Selection Theorems and their Applications. VII, 101 pages. 1972. DM 16,-

Vol. 264: W. Messing, The Crystals Associated to Barsotti-Tate Groups: With Applications to Abelian Schemes. III, 190 pages. 1972. DM 18,-

Vol. 265: N. Saavedra Rivano, Catégories Tannakiennes. II, 418 pages. 1972. DM 26,-

Vol. 266: Conference on Harmonic Analysis. Edited by D. Gulick and R. L. Lipsman. VI, 323 pages. 1972. DM 24,-

Vol. 267: Numerische Lösung nichtlinearer partieller Differential- und Integro-Differentialgleichungen. Herausgegeben von R. Ansorge und W. Törnig. VI, 339 Seiten. 1972. DM 26,-

Vol. 268: C. G. Simader, On Dirichlet's Boundary Value Problem. IV, 238 pages. 1972. DM 20,-

Vol. 269: Théorie des Topos et Cohomologie Etale des Schémas. (SGA 4). Dirigé par M. Artin, A. Grothendieck et J. L. Verdier. XIX, 525 pages. 1972. DM 50,-

Vol. 270: Théorie des Topos et Cohomologie Etale des Schémas. Tome 2. (SGA 4). Dirigé par M. Artin, A. Grothendieck et J. L. Verdier. V, 418 pages. 1972. DM 50,-

Vol. 271: J. P. May, The Geometry of Iterated Loop Spaces. IX, 175 pages. 1972. DM 18,-

Vol. 272: K. R. Parthasarathy and K. Schmidt, Positive Definite Kernels, Continuous Tensor Products, and Central Limit Theorems of Probability Theory. VI, 107 pages. 1972. DM 16,-

Vol. 273: U. Seip, Kompakt erzeugte Vektorräume und Analysis. IX, 119 Seiten. 1972. DM 16,-

Vol. 274: Toposes, Algebraic Geometry and Logic. Edited by. F. W. Lawvere. VI, 189 pages. 1972. DM 18,-

Vol. 275: Séminaire Pierre Lelong (Analyse) Année 1970-1971. VI, 181 pages. 1972. DM 18,-

Vol. 276: A. Borel, Représentations de Groupes Localement Compacts. V, 98 pages. 1972. DM 16,-

Vol. 277: Séminaire Banach. Edité par C. Houzel. VII, 229 pages. 1972. DM 20,-

Vol. 278: H. Jacquet, Automorphic Forms on GL(2). Part II. XIII, 142 pages. 1972. DM 16,-

Vol. 279: R. Bott, S. Gitler and I. M. James, Lectures on Algebraic and Differential Topology. V, 174 pages. 1972. DM 18,-

Vol. 280: Conference on the Theory of Ordinary and Partial Differential Equations. Edited by W. N. Everitt and B. D. Sleeman. XV, 367 pages. 1972. DM 26,-

Vol. 281: Coherence in Categories. Edited by S. Mac Lane. VII, 235 pages. 1972. DM 20,-

Vol. 282: W. Klingenberg und P. Flaschel, Riemannsche Hilbertmannigfaltigkeiten. Periodische Geodätische. VII, 211 Seiten. 1972. DM 20,-

Vol. 283: L. Illusie, Complexe Cotangent et Déformations II. VII, 304 pages. 1972. DM 24,-

Vol. 284: P. A. Meyer, Martingales and Stochastic Integrals I. VI, 89 pages. 1972. DM 16,-

Vol. 285: P. de la Harpe, Classical Banach-Lie Algebras and Banach-Lie Groups of Operators in Hilbert Space. III, 160 pages. 1972. DM 16,-

Vol. 286: S. Murakami, On Automorphisms of Siegel Domains. V, 95 pages. 1972. DM 16,-

Vol. 287: Hyperfunctions and Pseudo-Differential Equations. Edited by H. Komatsu. VII, 529 pages. 1973. DM 36,-

Vol. 288: Groupes de Monodromie en Géométrie Algébrique. (SGA 7 I). Dirigé par A. Grothendieck. IX, 523 pages. 1972. DM 50,-

Vol. 289: B. Fuglede, Finely Harmonic Functions. III, 188. 1972. DM 18,-

Vol. 290: D. B. Zagier, Equivariant Pontrjagin Classes and Applications to Orbit Spaces. IX, 130 pages. 1972. DM 16,-

Vol. 291: P. Orlik, Seifert Manifolds. VIII, 155 pages. 1972. DM 16,-

Vol. 292: W. D. Wallis, A. P. Street and J. S. Wallis, Combinatorics: Room Squares, Sum-Free Sets, Hadamard Matrices. V, 508 pages. 1972. DM 50,-

Vol. 293: R. A. DeVore, The Approximation of Continuous Functions by Positive Linear Operators. VIII, 289 pages. 1972. DM 24,-

Vol. 294: Stability of Stochastic Dynamical Systems. Edited by R. F. Curtain. IX, 332 pages. 1972. DM 26,-

Vol. 295: C. Dellacherie, Ensembles Analytiques, Capacités, Mesures de Hausdorff. XII, 123 pages. 1972. DM 16,-

Vol. 296: Probability and Information Theory II. Edited by M. Behara, K. Krickeberg and J. Wolfowitz. V, 223 pages. 1973. DM 20,-

Vol. 297: J. Garnett, Analytic Capacity and Measure. IV, 138 pages. 1972. DM 16,-

Vol. 298: Proceedings of the Second Conference on Compact Transformation Groups. Part 1. XIII, 453 pages. 1972. DM 32,-

Vol. 299: Proceedings of the Second Conference on Compact Transformation Groups. Part 2. XIV, 327 pages. 1972. DM 26,-

Vol. 300: P. Eymard, Moyennes Invariantes et Représentations Unitaires. II, 113 pages. 1972. DM 16,-

Vol. 301: F. Pittnauer, Vorlesungen über asymptotische Reihen. VI, 186 Seiten. 1972. DM 18,-

Vol. 302: M. Demazure, Lectures on p-Divisible Groups. V, 98 pages. 1972. DM 16,-

Vol. 303: Graph Theory and Applications. Edited by Y. Alavi, D. R. Lick and A. T. White. IX, 329 pages. 1972. DM 26,-

Vol. 304: A. K. Bousfield and D. M. Kan, Homotopy Limits, Completions and Localizations. V, 348 pages. 1972. DM 26,-

Vol. 305: Théorie des Topos et Cohomologie Etale des Schémas. Tome 3. (SGA 4). Dirigé par M. Artin, A. Grothendieck et J. L. Verdier. VI, 640 pages. 1973. DM 50,-

Vol. 306: H. Luckhardt, Extensional Gödel Functional Interpretation. VI, 161 pages. 1973. DM 18,-

Vol. 307: J. L. Bretagnolle, S. D. Chatterji et P.-A. Meyer, Ecole d'été de Probabilités: Processus Stochastiques. VI, 198 pages. 1973. DM 20,-

Vol. 308: D. Knutson, λ-Rings and the Representation Theory of the Symmetric Group. IV, 203 pages. 1973. DM 20,-

Vol. 309: D. H. Sattinger, Topics in Stability and Bifurcation Theory. VI, 190 pages. 1973. DM 18,-

Vol. 310: B. Iversen, Generic Local Structure of the Morphisms in Commutative Algebra. IV, 108 pages. 1973. DM 16,-

Vol. 311: Conference on Commutative Algebra. Edited by J. W. Brewer and E. A. Rutter. VII, 251 pages. 1973. DM 22,-

Vol. 312: Symposium on Ordinary Differential Equations. Edited by W. A. Harris, Jr. and Y. Sibuya. VIII, 204 pages. 1973. DM 22,-

Vol. 313: K. Jörgens and J. Weidmann, Spectral Properties of Hamiltonian Operators. III, 140 pages. 1973. DM 16,-

Vol. 314: M. Deuring, Lectures on the Theory of Algebraic Functions of One Variable. VI, 151 pages. 1973. DM 16,-

Vol. 315: K. Bichteler, Integration Theory (with Special Attention to Vector Measures). VI, 357 pages. 1973. DM 26,-

Vol. 316: Symposium on Non-Well-Posed Problems and Logarithmic Convexity. Edited by R. J. Knops. V, 176 pages. 1973. DM 18,-

Vol. 317: Séminaire Bourbaki – vol. 1971/72. Exposés 400–417. IV, 361 pages. 1973. DM 26,-

Vol. 318: Recent Advances in Topological Dynamics. Edited by A. Beck. VIII, 285 pages. 1973. DM 24,-

Vol. 319: Conference on Group Theory. Edited by R. W. Gatterdam and K. W. Weston. V, 188 pages. 1973. DM 18,-

Vol. 320: Modular Functions of One Variable I. Edited by W. Kuyk. V, 195 pages. 1973. DM 18,-

Vol. 321: Séminaire de Probabilités VII. Edité par P. A. Meyer. VI, 322 pages. 1973. DM 26,-

Vol. 322: Nonlinear Problems in the Physical Sciences and Biology. Edited by I. Stakgold, D. D. Joseph and D. H. Sattinger. VIII, 357 pages. 1973. DM 26,-

Vol. 323: J. L. Lions, Perturbations Singulières dans les Problèmes aux Limites et en Contrôle Optimal. XII, 645 pages. 1973. DM 42,-

Vol. 324: K. Kreith, Oscillation Theory. VI, 109 pages. 1973. DM 16,-

Vol. 325: Ch.-Ch. Chou, La Transformation de Fourier Complexe et L'Equation de Convolution. IX, 137 pages. 1973. DM 16,-

Vol. 326: A. Robert, Elliptic Curves. VIII, 264 pages. 1973. DM 22,-

Vol. 327: E. Matlis, 1-Dimensional Cohen-Macaulay Rings. XII, 157 pages. 1973. DM 18,-

Vol. 328: J. R. Büchi and D. Siefkes, The Monadic Second Order Theory of All Countable Ordinals. VI, 217 pages. 1973. DM 20,-

Vol. 329: W. Trebels, Multipliers for (C, α)-Bounded Fourier Expansions in Banach Spaces and Approximation Theory. VII, 103 pages. 1973. DM 16,-

Vol. 330: Proceedings of the Second Japan-USSR Symposium on Probability Theory. Edited by G. Maruyama and Yu. V. Prokhorov. VI, 550 pages. 1973. DM 36,-

Vol. 331: Summer School on Topological Vector Spaces. Edited by L. Waelbroeck. VI, 226 pages. 1973. DM 20,-

Vol. 332: Séminaire Pierre Lelong (Analyse) Année 1971-1972. V, 131 pages. 1973. DM 16,-

Vol. 333: Numerische, insbesondere approximationstheoretische Behandlung von Funktionalgleichungen. Herausgegeben von R. Ansorge und W. Törnig. VI, 296 Seiten. 1973. DM 24,-

Vol. 334: F. Schweiger, The Metrical Theory of Jacobi-Perron Algorithm. V, 111 pages. 1973. DM 16,-

Vol. 335: H. Huck, R. Roitzsch, U. Simon, W. Vortisch, R. Walden, B. Wegner und W. Wendland, Beweismethoden der Differentialgeometrie im Großen. IX, 159 Seiten. 1973. DM 18,-

Vol. 336: L'Analyse Harmonique dans le Domaine Complexe. Edité par E. J. Akutowicz. VIII, 169 pages. 1973. DM 18,-

Vol. 337: Cambridge Summer School in Mathematical Logic. Edited by A. R. D. Mathias and H. Rogers. IX, 660 pages. 1973. DM 42,-

Vol. 338: J. Lindenstrauss and L. Tzafriri, Classical Banach Spaces. IX, 243 pages. 1973. DM 22,-

Vol. 339: G. Kempf, F. Knudsen, D. Mumford and B. Saint-Donat, Toroidal Embeddings I. VIII, 209 pages. 1973. DM 20,-

Vol. 340: Groupes de Monodromie en Géométrie Algébrique. (SGA 7 II). Par P. Deligne et N. Katz. X, 438 pages. 1973. DM 40,-

Vol. 341: Algebraic K-Theory I, Higher K-Theories. Edited by H. Bass. XV, 335 pages. 1973. DM 26,-

Vol. 342: Algebraic K-Theory II, "Classical" Algebraic K-Theory, and Connections with Arithmetic. Edited by H. Bass. XV, 527 pages. 1973. DM 36,-